YO-ALL-428

ENGINEERING

BOARD OF ACADEMIC EDITORIAL ADVISERS

Professor Joseph Bram, *Anthropology*
Professor Harry A. Charipper, *Biology and Zoology*
Professor Isidor Chein, *Psychology*
Professor Hollis R. Cooley, *Mathematics*
Associate Professor William J. Crotty, *Botany*
Professor Jotham Johnson, *Archaeology*
Professor Serge A. Korff, *Astronomy*
Professor James E. Miller, *Meteorology*
Professor George M. Murphy, *Chemistry; Science*
Professor Gerhard Neumann, *Oceanography*
Professor Joseph Pick, M.D., *Anatomy and Physiology*
Dean John R. Ragazzini, *Engineering*
Professor Morris H. Shamos, *Physics*
Professor Leslie E. Spock, *Geology*

THE NEW YORK UNIVERSITY LIBRARY OF SCIENCE

ENGINEERING

EDITED BY

Samuel Rapport AND *Helen Wright*

ACADEMIC EDITORIAL ADVISER

JOHN R. RAGAZZINI

Dean of the College of Engineering and Professor of Electrical Engineering, New York University

NEW YORK UNIVERSITY PRESS 1963

© 1963 by New York University
Library of Congress Catalog Card Number: 62–18019
Manufactured in the United States of America
Designed by Andor Braun

ACKNOWLEDGMENTS

"Early Construction Methods" from *What Engineers Do* by Walter Binger. Copyright 1928, revised edition copyright 1938, by W. W. Norton & Company, Inc. Reprinted by permission of W. W. Norton & Company, Inc.

"The Growth of Mechanical Power" from *The Growth of Mechanical Power* by Miles Tomalin. Reprinted by permission of Methuen & Company, Ltd.

"Leonardo da Vinci, Military Engineer" by Bern Dibner. From *Studies and Essays in the History of Science and Learning Offered in Homage to George Sarton* edited by M. F. Ashley Montagu. Reprinted by permission of Abelard-Schuman Limited.

"James Watt" from *James Watt* by Ivor Hart. Copyright 1949 by Henry Schuman, Inc. Reprinted by permission of Abelard-Schuman Limited.

"The Steam Turbine and Other Inventions of Sir Charles Parsons" by R. H. Parsons. Reprinted by permission of the author.

"From Faraday to the Dynamo" by Harold I. Sharlin, originally published in *Scientific American*, May, 1954. Copyright © 1954 by Scientific American, Inc. and reprinted by their permission. All rights reserved.

"The Internal Combustion Engine" from *A Short History of Technology* by T. K. Derry and T. I. Williams. Reprinted by permission of The Clarendon Press.

"Reaction Motors" from *The Coming Age of Rocket Power* by G. Edward Pendray. Copyright 1945, 1947 by Harper & Brothers. Reprinted by permission of Harper & Brothers.

"The A.B.C. of Atomic Energy" by Sir Christopher Hinton. Reprinted by permission of The British Broadcasting Corporation.

"The Age of Steel" from *Science and Industry in the Nineteenth Century* by J. D. Bernal. Reprinted by permission of Routledge Kegan Paul Ltd.

"Manufacture of Wire Ropes" by John A. Roebling from the *American Railroad Journal* and *Mechanics Magazine*.

"Discovering, Recovering and Concentrating the Minerals" from *The Storehouse of Civilization* by C. C. Furnas. Copyright 1939 by Teachers College, Columbia University. Reprinted by permission of the Bureau of Publications, Teachers College, Columbia University.

"Tools and Automation" from *Automation and Social Progress* by S. Lilley. Reprinted by permission of Lawrence and Wishart, Ltd.

"Skyscraper: Skin for its Bones" from *Architecture and the Esthetics of Plenty* by James Marston Fitch. Copyright © 1961 by Columbia University Press. Reprinted by permission of Columbia University Press.

"Building the Union Pacific Railroad" from *Trails, Rails and War* by J. R. Perkins. Copyright 1929 by The Bobbs-Merrill Company, © 1956 by J. R. Perkins. Reprinted by special permission of the publishers, The Bobbs-Merrill Company, Inc.

"Down in the Caissons" from *The Builders of the Bridge* by D. B. Steinman. Copyright, 1945, 1950 by Harcourt, Brace & World, Inc., and reprinted by their permission.

"Tesla and Alternating Current" from *Prodigal Genius* by John J. O'Neill. Copyright, 1944 by Ives Washburn, Inc., and reprinted by their permission.

"Triumph at Messines" from *In Flanders Fields* by Leon Wolff. Copyright © 1958 by Leon Wolff. Reprinted by permission of The Viking Press, Inc.

"Abu Simbel" by Ritchie Calder. Reprinted from *The Unesco Courier*.

"Atlantropa—the Changed Mediterranean" from *Engineers' Dreams* by Willy Ley. Copyright 1954 by Willy Ley. Reprinted by permission of The Viking Press, Inc.

"Outer Space: The Technical Prospects" by H. Guyford Stever, from *Outer Space: Prospects for Man and Society,* by The American Assembly, Columbia University, New York. Copyright © 1962 by The American Assembly, Columbia University. Reprinted by permission of Prentice-Hall, Inc., publisher.

CONTENTS

FOREWORD ix

INTRODUCTION xi

I. The Beginnings

Early Construction Methods *Walter D. Binger* 3
The Growth of Mechanical Power 11
  Miles Tomalin
Leonardo da Vinci: Military Engineer 30
  Bern Dibner

II. The Age of the Machine

A. *Power*

James Watt and the History of Steam Power 44
  Ivor B. Hart
Sir Charles Parsons and the Steam Turbine 64
  R. H. Parsons
From Faraday to the Dynamo 82
  Harold I. Sharlin
The Internal Combustion Engine 95
  T. K. Derry AND *Trevor I. Williams*
Reaction Motors: How They Work and Some 105
Experiments With Them
  G. Edward Pendray
The A.B.C. of Atomic Energy 126
  Sir Christopher Hinton

B. *Materials and Methods*

The Age of Steel *J. D. Bernal* 145
Manufacture of Wire Ropes 157
  John A. Roebling
Discovering, Recovering, and Concentrating 164
  the Minerals *C. C. Furnas*

Tools and Automation *Samuel Lilley* 181
Skyscraper: Skin for Its Bones 221
 James Marston Fitch

III. Episodes in Engineering History

Building the Union Pacific Railroad 237
 J. R. Perkins
The Caissons of the Brooklyn Bridge 255
 D. B. Steinman
Tesla and Alternating Current 297
 John J. O'Neill
Triumph at Messines *Leon Wolff* 313

IV. Toward the Future

Abu Simbel *Ritchie Calder* 328
Atlantropa—The Changed Mediterranean 338
 Willy Ley
Outer Space—The Technical Prospects 351
 H. Guyford Stever

FOREWORD

IT HAS been asserted by C. P. Snow, the eminent scientist and author, that there are two discernible cultures in our modern society, that of the humanist and that of the scientist and engineer. He emphasizes that there is insufficient communication between these two cultures brought about, to large degree, by the differences in language, content, and terminology. To find a mode of expression to describe so complex a field as engineering, which is both understandable to the layman not trained in engineering yet acceptable to the engineer as being sufficiently accurate, is no small task. This objective was set in this book and the introductions to the numerous articles are designed to set the stage for what follows. The authors of the articles and editors of this volume have succeeded admirably in maintaining technical accuracy while expressing themselves in simple, forthright language.

Through the ages, engineering has been a moving force in the shaping of our society in peace and war. From the early beginnings through the age of power, materials, and methods, culminating in the important episodes in engineering history and with a speculative eye to the future, this book traces the development of engineering. Because of the enormous advances in the field during the past century, it is impossible for any one individual to acquire competence in more than one or two disciplines and, for this reason, many eminent engineers, scientists, and engineering historians have been included, with articles in their respective specialties.

The articles give the layman an understanding of what the

engineering profession is, what it does, and how it affects the society of which it is a part. Engineering is one of the most potent factors affecting the course of modern history. If culture is defined as an understanding of civilization, the engineering profession can hardly be omitted, and this volume may be looked upon as a cultural contribution. I recommend it both to layman and to my colleagues in the engineering profession. The former will derive pleasure and information while the latter will better understand the glorious background which has led them to their present state of technical achievement and their relation to the society which they serve.

John R. Ragazzini

INTRODUCTION

A GLANCE at any standard dictionary reveals the huge range of activity with which the engineer is involved. Figuratively speaking, he "carries through any scheme or enterprise by skill or artful management." More exactly, he is one "who contrives, plans or invents," but this all-inclusive definition can be subdivided into literally dozens of categories. The military engineer is "a designer and constructor of military works"; the civil engineer is versed in "the design, construction and care of roads, bridges, canals, aqueducts, etc."; the mechanical engineer concerns himself with "engines and machines." There are electrical, mining, chemical, and structural engineers; there are management and manufacturing engineers. The number of specialties has been compounded by the almost inconceivable complexity of modern civilization.

Far-ranging as these definitions may appear, they share one thing in common. The primary aim of the engineer is to get things done in the most efficient, most economical manner possible. The profession concerns itself with tools that can do things better than the human hand; with power which can outdo the strength of muscles a billionfold; with materials whether they occur in nature or are contrived by the physicist or chemist; above all with methods for using these components to any desired end.

The profession is nearly as old as the race of *homo sapiens*. The Paleolithic man who contrived a rude flint hatchet was an engineer. His descendant who used it to fell a tree across a stream was a more sophisticated example of the breed. And *his* descendants—who discovered the wheel, the plow, the use of iron, the catapult in war—are the lineal forefathers of those who design atomic-power plants and hurl rockets at the moon.

The engineer has an ancestry far older than the basic scientist. He is concerned not so much with unraveling fundamental laws of nature as with bending them to his own purposes. The South American savage knows nothing of the laws of either gravitation or catenaries, but by a process of trial and error he succeeds in building a vine suspension bridge which carries him safely across a gorge. The Neolithic builder of a hut who pries loose a stone with the branch of a tree is unaware that he is employing the principle of the lever. For thousands of years, the engineer created massive structures, irrigated his land, and invented weapons to slay his food and his enemies with only the most rudimentary knowledge of the scientific laws involved; yet to this day many of the works of his hands evoke a feeling of almost incredulous awe.

In more sophisticated times, as scientific knowledge has increased, the engineer has eagerly grasped at the clues it offers. On occasion, as in the case of the electronics engineer, he has made a considerable contribution to the fund of basic knowledge. But while he works in ever closer collaboration with the chemist, the physicist, or the geologist, and while many of his profession use only pencil and paper as tools, he remains in essence a maker, a doer, a man of action. He is the solver of specific problems—the link between the theory and the end result. An Einstein discovers the relationship between energy and mass. The engineer builds the equipment which puts the equation to work. Today he is concerned not only with such relatively prosaic matters as the building of superhighways and skyscrapers but also with ever newer methods of harnessing the power in the sun's rays or devising materials which will withstand extremes of temperature and shock.

Unlike such sciences as physics and astronomy, the history of engineering is not studded with the names of outstanding geniuses. A Watt, a Parsons, or a Roebling have had great influence on the development of the profession, but in general engineering has grown through a slow accretion of knowledge—through innumerable small contributions by many practical men. Yet in the mass, the effect of these contributions has been so great that modern civilization would be unthinkable without them.

Above all, the calling of the engineer is fraught with glamour. The ability to cope with savage nature is part of his stock in trade. In the most literal sense the engineer makes the desert bloom and the bowels of the earth give up their treasures. The day will soon arrive when, with his companion worker the pure scientist, he will provide the power and construct the machines which, if mankind can learn to live with them, will eliminate the drudgery with which he has been burdened since Adam.

Any attempt even to touch on all phases of the science and profession of engineering in a volume of this size is obviously foredoomed to failure. It is hoped, however, that in the selections that follow, the reader will obtain some idea of the problems the engineer has faced and how he has gone about solving them. Because of space limitations, some of these selections are presented in abbreviated form; and as the book is intended for the educated laymen rather than for the student and specialist, scholarly apparatus has in general been omitted.

This is one of the books in the series entitled THE NEW YORK UNIVERSITY LIBRARY OF SCIENCE. Other volumes, dealing with individual sciences, are published and in preparation. The series as a whole will encompass much of the universe of modern man, for that universe has been shaped in greatest measure by science, the branch of human activity whose name is derived from the Latin *scire*—to know.

ENGINEERING

I. The Beginnings

I. The Beginnings

The men of the Stone Age fulfilled their primitive wants with materials of stone and wood, tools of flint or horn, and human muscle power. Their primary needs were weapons for hunting and war and shelter from the elements, and their engineering accomplishments were small. They knew nothing of chemistry, metallurgy or agriculture. They must have observed the power of water in motion but had no glimmering of how to harness it. They were to wait tens of thousands of years before even discovering that animal power could be substituted for that of human beings.

Gradually man learned that permanent structures were more satisfactory than tents of hide; that animal husbandry and agriculture supplied food more efficiently than hunting deer. Some time around the sixth millenium B.C. and somewhere in the Middle East—probably Mesopotamia—he began to cluster in tiny settlements. As the settlements grew into villages and then into cities, specialization arose. Priests and witch doctors, herders of sheep and tillers of the soil took on special functions. Among these specialists was the artisan. By trial and error he gradually amassed a body of knowledge which he used to build dwellings and temples and to dig canals to water his fields. He had already learned to grind and polish. With agriculture, his tools became more complicated. He domesticated the horse. He first learned to use copper, then bronze and finally iron. His efficiency, during these opening days of civilization, increased manyfold and with it came its invariable concomitant—what would now be called a population explosion.

City-states developed, conquered their neighbors, and became nations. A great center of civilization arose in the neighborhood of the Tigris and Euphrates—the territory which the American historian James H. Breasted named "the fertile crescent." The builders constructed palaces and temples which have now crumbled. They constructed embankments and levies to contain flood waters. A huge dam—the Marib—is

1

The Beginnings

said to have supported a prosperous agricultural community near the shores of the Red Sea; an aqueduct supplied water to the city of Nineveh; roads and bridges cemented the empires of Xerxes and Darius.

In Egypt, where civilization existed for millenia before the birth of Christ, the "kings' builders" left more impressive and permanent monuments because they worked in stone. They hewed temples like that at Abu Simbel out of solid rock. They constructed pyramids from blocks weighing many tons, used oared barges to carry construction material, and channeled the waters of the Nile—all with mathematical exactness and precision.

These two Middle Eastern civilizations were followed by those of Greece and Rome. Perhaps never before or since have engineering and art been so ideally wedded as in the architecton—the master builder of Greece. His creations had a matchless grace which has survived the depredations of the centuries. The Romans, with their roads, their arched bridges, their public buildings, their aqueducts and tunnels, were the architecti—the civil and military engineers par excellence of ancient times. Their point of view is aptly expressed in a statement by Frontinus, water commissioner of Rome, who wrote, "Will anybody compare the idle Pyramids, or those other useless though renowned works of the Greeks, with these aqueducts, with these many indispensable structures?"

To further the arts of war and peace, numerous ingenious tools and weapons were gradually evolved, and the Latins used the word ingenium to describe them. Over the centuries, the concept of the "engineer," the "ingénieur," the "engignier," the "ingeniarius"—the specialist in the contrivance of such machines—arose. His profession merged with that of the master builder to become the engineering of modern times.

In the Middle Ages, the engineer's most spectacular achievement was the construction of the great cathedrals. He also developed the use of wind and water power, and designed and constructed sound and beautiful bridges. In the Renaissance, he worked with more sophistication. But even in that period, and for several centuries thereafter, there was no sharp and drastic change in his point of view. The methods and materials he used were primarily those of his predecessors. In this article, taken from What Engineers Do, we are offered a glimpse of how such materials and methods influenced his craft.

Walter D. Binger, the author, is a construction engineer who, as Commissioner of the Borough Works of Manhattan, directed the design and construction of the East River Drive and Harlem River Drive in New York City.

EARLY CONSTRUCTION METHODS

WALTER D. BINGER

THE HISTORY OF MAN may be read by means of his tools and structures. From the earliest prehistoric periods his degree of culture may, through these relics, be identified and compared not only with that of his contemporaries of other races but with that of men of earlier and later times.

Long before man built any permanent structures for dwellings he had perfected certain tools required for his daily needs. In Europe he lived in caves during what is called the Paleolithic Period, or Old Stone Age, but his tools, such as flint knives and axes, were already very good.

In the United States the northern Indians, living in a stone age very like that which existed thousands of years earlier in Europe, made their dwellings in tents which have vanished, but they have left a record of flint arrowheads, axes, and other tools and weapons.

The development of structures in all countries has, of course, been based at first on the kind of building materials most conveniently available. If a man lived in a country with many natural caves he would have no need to think of using animal hides for the roof and walls of his house. If he lived in a thickly forested land, he would have little reason to try to cut stone.

The earliest Egyptian dwellings were of wood frame, or skeleton construction. Since the timber of the country furnishes only small beams, the walls and roofs of these houses were made of reeds, lotus, and papyrus. Not until their needs led the Egyptians to require larger and more permanent buildings did they take the next step in construction—building with mud. This they mixed with straw and rammed between molds, or forms, much as we now do with concrete, removing the boards as soon as the mud had hardened into walls, straight or curved as desired. The mud walls had to be thick to support the roof, or even to carry themselves, but thickness would not help if the walls were uniform from top to bottom. Since the mud could stand only a limited weight, lightness had to be secured by making the wall thinner at the top than at the bottom.

About this time they also started building with sunbaked bricks made from the same mud. Being a race of natural builders they carried each of these crude materials to a high state of perfection before they abandoned it for the next more difficult one. In brick work as well as stone, in which they later achieved their great glory, they soon discovered many ways of building by which the material could be made to express their ideas of ornamentation as well as fulfill their structural requirements.

All of these methods have been rediscovered hundreds of times and some of them were in use about a thousand years before by other peoples. For example, the Egyptians of this period, more than 3000 B.C, used the corbel, a system of making each higher course of bricks overlap the one below it. By placing two such columns as are shown in the figure in a position so that the topmost bricks touched, the result was a pointed arch. It is, however, a false arch since one side is independent of the other. The false arches of the Egyptians are probably not the oldest for they were used in brick sewers found in Nineveh which are said to date from 4000 B.C.

The Egyptians, driven on by their religious requirements, progressed from mud to the use of stone—limestone, sandstone, and granite, carefully quarried and cut. As their tools developed they produced works in stone of such size and magnificence as the world has received from no other peo-

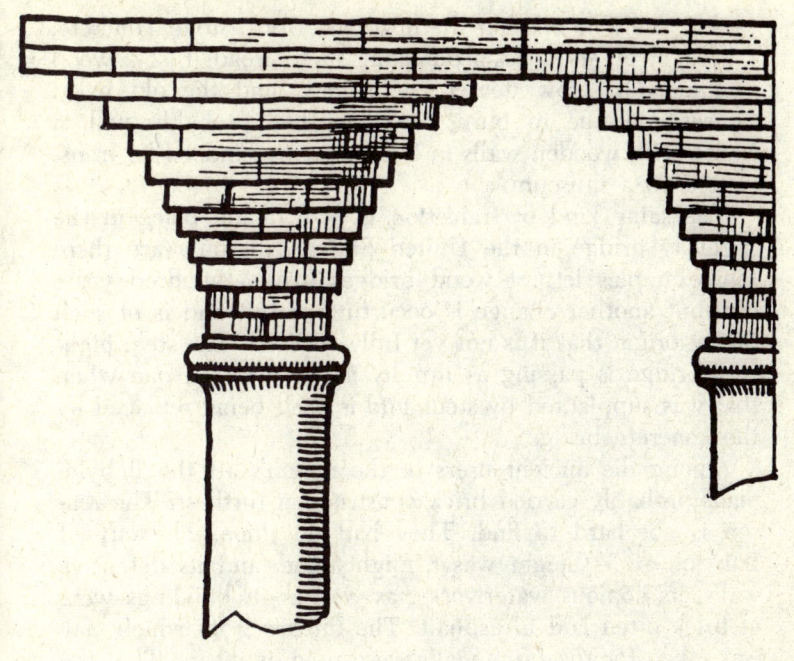

I-1. *A false arch.*

ple. They retained the same slope of walls required in their early mud temples, when they built in stone that might have been laid perpendicularly. As is well known, they also preserved, as ornaments in stone, the leaves and reeds which had earlier formed a structural portion of the skeleton frame houses. The same clinging to former shapes may be seen in the similarity of our earlier motor cars to the horse-drawn vehicles they succeeded. The buttons on the sleeves of our coats are also an example of the retention as ornament of something which formerly had a use. Today we know the Egyptians only for their work in stone, and the world in general has forgotten what led up to it.

New England, too, having at first the choice of timber from its forests and stone from its fields, chose the former and became a country of wood dwellings and buildings. But who in a thousand years from now will know it as such? Surely no one but the historian. Brick, stone, concrete, stucco, steel, and glass are making such inroads that if wood continues to grow dearer, as it may, and the old wood houses continue to burn down as they probably will, a house with wooden walls in 3000 A.D. may indeed be transported to a museum.

The same kind of transition is now taking place in the highway bridge in the United States. Not only are there few examples left of wood bridges over a hundred years old, but another change is occurring so fast and is of such recent origin that it is not yet fully realized. The steel highway bridge is passing as rapidly as the wooden one when that was supplanted by steel, and is itself being replaced by the concrete bridge.

Among the ancient users of the mason's art the Babylonians probably carried brick construction furthest. The reason is not hard to find. They had no stone. The city of Babylon at its height was a mighty one, and its defensive walls, its copious waterworks, as well as its buildings were of brick often laid in asphalt. The ditches from which clay was taken for the brick walls were used as moats. That the bricks were of good quality cannot be doubted, for many of those made in Biblical times have again found their way into buildings constructed in our eighteenth and nineteenth centuries. One of these bricks bears the stamp: "Nebuchadnezzar, King of Babylon, restorer of the Pyramid and the Tower, eldest son of Nabopolassar, King of Babylon."

Undoubtedly the question of greatest interest to the engineer is how the ancient builders of huge stone structures achieved them without any power other than that generated by the bodies of men and animals. The magnitude of these works was so enormous that one at any rate still stands as the largest structure ever erected by man. It is the Great Wall of China which was begun in the third century B.C. Fifteen hundred miles long and about 50 feet high in many places, it follows mountains and valleys and seems almost to be part of the natural landscape. Arches were employed to carry the wall across streams and valleys.

The builders of those ancient structures had pulleys,

windlasses, and capstans. The two latter are instruments on which rope may be wound up and heavy weights lifted, just as one winds up a fish line on a reel against the pull of the fish, or an anchor on its chain. They knew of the force that could be exerted by levers and are believed to have made much use of them in raising large stones. They realized that almost unbelievable loads could be pulled on rollers, by enough men, even up an incline.

The Egyptians must have used all of these implements in erecting the Great Pyramid. This, their most impressive monument, is 764 feet square and 450 feet high and is built of granite and limestone some of which was brought from quarries 500 miles up the Nile.

Given ropes and pulleys and enough men, there is no reason why a 50-story skyscraper could not be built entirely without power machinery. Of course, machinery would be required to produce the structural steel members and many other materials such as cold drawn steel wire, steel and brass pipe, heating, plumbing, and lighting fixtures, but in that we are not interested. The whole building could be erected by manpower as the Egyptians erected their tombs. Instead of finishing a floor in a day it would probably take a year, but every I-beam could be hauled from Pittsburgh to New York by a team of horses, hoisted into place by 100 men at the end of a rope, and riveted by blacksmiths using hand hammers. It is well to bear these things in mind so as not to be lost in wonder, instead of admiration, when we contemplate the work of the Egyptians. We need not be sparing with our admiration for they showed ingenuity, organizing ability, and a determination to succeed over obstacles which would have seemed insuperable to a less resourceful race of builders.

The ancients were undoubtedly supreme in the use of the masonry of great stones, and the Egyptians were probably the ones who introduced that use to the rest of the world. Eventually the building of structures of great stones, megalithic masonry, or heliolithic as it is sometimes called, found its way all around the earth, eastward to India and China, westward to Brittany where it appears in tombs, and to the British Isles where it is found in the stones of many tons in Stonehenge, and to a lesser extent to the New World, where the Egyptian style reappears in Mexico and Yucatan. Of course, it is not actually known whether it is simply by chance that similar architectural types occur in such widely separated places as Egypt and Yucatan at such different times, or whether the pyramids of the latter were inspired by those of the former. Much is being learned every year, however, and it is an absorbing subject for the engineer whose interest is aroused by it.

With the Romans there appears the first consistent and common use of concrete made from hydraulic cement, which will harden under water, a use which foreshadows that of the concrete age. In their waterworks, harbor developments, and foundations, as well as in their arches and domed ceilings, they made use of concrete poured into forms, in large quantities.

Instead of attempting to excel in piling up masses of huge stones, the Romans prided themselves on buildings in which genius of design rather than of construction is seen. It was as though the scientific age of the Greeks had blossomed again in the brilliant technical era of the Roman engineers. But while there are few basic forms of structure in use today that were unknown to them, they had apparently made practically no advance over the Greeks in the fundamental studies of physics, hydraulics, and mathematical surveying, on which these structures depended. Because of this they had been unable to make revolutionary improvements over the Greek machinery, tools, and instruments, being limited to Greek principles.

Their use of arch ribs such as constitute the roof of the Pantheon in Rome is an example of technical achievement which has perhaps never been surpassed. The ribs were made of Roman bricks on edge. They were flat, about a foot square, and more like our tiles than our bricks. Centering, or supports, carried the ribs, the filling being placed upon them. This consisted of alternate layers of the tiles laid flat, and of concrete, until a great thickness had been secured. The entire dome was self-supporting and rigid enough not to thrust out the walls, though they had been made solid and 20 feet thick to resist any such tendency.

In the fifteenth century Brunelleschi, a great student of the Roman style, built a dome 130 feet in diameter and 135 feet high on the cathedral in Florence. He did this without centering by a method that seems to have been original with him. The dome consisted of a series of horizontal rings each a little smaller than that below it. In effect it was the ancient corbel system reappearing in a very new and brilliant use.

In an entirely different field, though one reaching back to earliest times, the Eddystone Lighthouse stands out in conception and execution. Built on the wave-swept site of two earlier towers it was completed in 1759 by John Smeaton, its designer. Its walls curved inwards as they rose, to lessen resistance to waves. Among other innovations was the scheme of having some stones dovetailed to the stones above, below, and on either side of them. Here we see the attempt of the builder in stone to make his structure monolithic. No wonder that modern lighthouses are built of concrete, reinforced with steel, which really is monolithic. Owing to the undermining of the foundation rock and the need for a higher tower the Eddystone Lighthouse was taken down and replaced in 1877. All subsequent lighthouses, however, have shown the effect of Smeaton's principles of design.

Today, structural steel has changed everything. Domes of the types in Rome and Florence have steel frames. Brick and stone, instead of being vital to the structure, are supported by the steel and act merely to keep out the wind and rain. They are more and more being replaced by glass.

The constant use of a material—wood, stone, concrete, or steel—for structural purposes, develops into an architect and engineer the workman who has begun to think in terms of that material. We have had examples of this in very recent times. About a hundred years ago when we entered the era of long-span wooden bridges, the finest and longest were built in Switzerland by a village carpenter who, from an engineering point of view, was uneducated. The Swiss since immemorial times had been accustomed to handle big timbers because of their forests of magnificent trees, and they had workmen who were so at home with this material that it was perfectly natural for one of them to make a great advance in the use of it. Thus the Swiss carpenter designed, among others, a bridge which for a long time carried heavy loads on spans of over 200 feet.

It is, of course, impossible to trace the history of building materials fully here, for an entire book might well be devoted to this one subject. And what a book it would be! Part of the history of the human race from its dawn down to this very minute might be written around man's longing and search for new and different materials with which to build his tombs, cathedrals, palaces, or skyscrapers.

Before the revolution in mechanical power, the most spectacular engineering achievements were in the field of construction, and the great achievements of civil engineers approximated those of our own day. There is far greater dissimilarity between the past and present of power production, yet it would be a mistake to assume that the science remained static before the great days of the steam engine. One of the most important human achievements of the Middle Ages was the gradual elimination of slavery through the substitution of nonhuman power. This power drew on three prime sources —wind, water, and the domesticated horse. Two relatively unsung inventions were the horseshoe and the horse collar, which resulted not only in the reduction of human drudgery but also in a new and more elegant form of transportation. The undershot water wheel which utilized the motion of water and the overshot wheel, depending on its weight, were joined in the eighteenth century by the breast wheel in which the water was directed against the blades nearly halfway up, utilizing both weight and motion. The power of the wind was utilized in Europe to turn windmills as early as the end of the twelfth century, and in the next three centuries more efficient sails and bonnet windmills—in which only the top was adjusted to the direction of the wind—were invented.

Increasingly large and complex machines turned grindstones, operated saws, pumped the bellows of forges and lifted water. Cranks and gears were devised to transfer power from horizontal to vertical axles. Although the invention of gunpowder revolutionized the strategy and tactics of war, peaceful utilization of the direct thrust of an explosion eluded mankind for hundreds of years. One of the most useful forms of power —resulting from the combination of water and heat—was of course observable in every kitchen. It was harnessed stumblingly, slowly, and indirectly. This article describes the first hesitant steps.

THE GROWTH OF MECHANICAL POWER

MILES TOMALIN

ALL HUMAN EFFORT begins with the energy of living muscles. Compared with other sources of power, the human body is weak and not very reliable; yet with his little muscle-engine, rating at about one-tenth of a horsepower, man can control forces vast enough to destroy whole cities in a fraction of a second, and lives in fear that in a fit of malice he will one day do so.

Man's first power-producing "machine" was some other living creature. He captured a wild beast, and with his superior brain and strength of will trained it to haul his loads for him, and carry him from place to place on its back. He fueled his horse-engine, as he fueled himself, with food.

In this fashion, man, primitive as he was, took a step ahead of all other animals. Small beasts might attach themselves to the bodies of larger ones and be carried about, but man alone had the wit to control the movement of his mount. His achievement was twofold. He had discovered a source of power; after that, he had invented a means of putting it to use. Discovery gave rise to invention: that has always been the order of things.

In the early days of tribal life, a tribe that was victorious in battle put the male captives to death. This, as they presently realised, was a wasteful procedure. If they spared their captives' lives and put them to work instead of to death, they themselves could live an easier and pleasanter life. For all its drawbacks, slavery was more humane than slaughter.

A human slave had not the strength of a horse, but he was cleverer, and could be set to do many things that horses were incapable of. The problem of weakness could be overcome by gathering many slaves into a gang and combining their strength.

According to the Greek historian Herodotus, the building of the Great Pyramid of Cheops occupied the labour of 100,000 slaves for twenty years. In the course of that time, the slave gangs set six million tons of stone in place. These gangs, we may assume, were the most powerful and efficient power engines of that period, which dates back some five or six thousand years.

The measurements of the Great Pyramid were calculated exactly, and closely kept. The men in charge of the building must have been skilful mechanics: mere strength could never have done it. To raise the huge stone blocks to the required height, a long sloping ramp was built, up which the blocks were hauled on rollers. The Egyptians knew how to make good use of the mechanical principles of the wheel and the inclined plane, and, of course, of the lever. The experience of the work itself must have taught them many new things. The efficiency of the gang-engine was increased by mechanical skill: skill increased because the gang-engine had need of it.

Another use for the slave gang of ancient days was to row Egyptian Pharaohs up and down the Nile in their broad galleys. According to inscriptions of the time, about twenty-five men on each side manned a bank of oars, or of paddles, so that one could say the boat was powered by a 50-man-power muscle engine.

Of all the natural sources of power, the strength of living muscles was the most obvious to simple men; but nature had put two others close under their noses, the flow of water and the rush of wind. It was not surprising that early use was made of them.

When at some very remote stage men first pushed logs out into the center of a river and were carried on them down the stream, they were harnessing the power of water and making it serve their uses. Many thousands of years later, they discovered that they could stretch a sail to the wind and be blown against and across the current. Some of the early Nile boats carried mast and sail to help the work of the oarsmen.

As time went by, mast and sail grew bigger and did more of the work. Adventurous merchants found it easier and quicker to carry goods from one country to another across the sea than to make the long trek round by land. The more they could dispense with oarsmen and their stores of food and water, the more room was left for goods. Sea trading led to the development of sails and rigging big enough to carry the ships along without the need of oarsmen at all on the open sea.

To put the power of wind and water to work on land needed more thought. Unlike a boat, a land engine using these moving elements must remain in place while the wind or the water moves past it, turning a wheel on its way.

Primitive water-wheels were grinding corn and raising water over river banks to irrigate the fields before the beginning of Christianity. The first corn-grinding mills were turned by a bladed wheel set flatwise in the stream, and so arranged that the stream flowed against one side of it only. The mill wheel and the upper millstone were fixed to a single shaft, and turned together, the shaft passing freely through a hole in the lower millstone. The lower stone stood still while the upper stone ground against it.

Vitruvius, a Roman architect living during the first century B.C., describes an improved form of water mill. His mill wheel was set upright, all but its lowest blades being free of the water. The current pressed against the submerged blades and moved the wheel round. The millstones had still to be set flatwise, however, or the grain would have run out before it was ground. This meant that the movement of the wheel had somehow to be transmitted through a right angle. It was one of the earliest known instances when the transmission of power appeared as a special problem. Vitruvius solved it by devising crude wooden gears of the kind known as "cog and lantern" wheels.

To spread a sail to drive a ship is so simple an idea that it may not even have occurred to the first sailors that the wind's power was being transmitted to the hull through the mast. Mounting a millstone on one shaft with the waterwheel presented no difficulty either. The time was to come, though much later, when the problem of transmission was to dominate the whole power story.

So long as men are foolish enough to go to war with one another, they must be clever enough to win their wars, if they can. For this reason, war has always been a stimulus to invention. In some early century B.C., military enthusiasts discovered that they could use the elasticity of cords or supple wood to hurl missiles at opposing troops, and the catapult was invented. The principle of elasticity as a source of power is in use all round us today, in the metal springs of clocks and watches.

When in the path of all-conquering Alexander the Great (fourth century B.C.) the center of learning and scholarship shifted from Athens to the new city of Alexandria, the priests in the Greek temples of that city became the chief students of the mechanical sciences. At some date within a century on either side of the birth of Christ, one of their number, called Hero, or Heron, compiled a book in which he described the construction and working of a variety of clever devices. Most of these produced magical-seeming effects which ignorant people would believe to come directly from the all-powerful gods. This would suit the needs of rulers who depended upon the obedience and subjection of their people.

It is unlikely that these inventions were all Hero's own: his book is probably a collection of all the ingenious mechanisms invented by his fellow priests in the years prior to his time. They show a knowledge of physics and mechanics that was not equalled in the West for more than a thousand years afterwards.

Here we find machines for opening the temple gates as if by an unseen hand: metal images which uttered unearthly sounds when the sun shone on them: a machine that automatically poured out a libation of wine when a sacrificial fire was lit at its foot: a little whizzing engine, called an "æolipile," that was spun round by the reaction of two jets of steam: even a coin-in-slot machine. Many of these inventions relied upon the simple fact that liquids and gases expand when heated, and must therefore force their way out of the vessel that contains them by whatever passage is left open. The whistling statue was hollow: the air, expanding in the heat of the sun, escaped through its mouth, which was fashioned like the sounding part of an organ pipe. In the case of the gate-opening mechanism, steam forming in a boiler pressed the water out through a spout, beneath which a bucket was suspended. The weight of the water caused the bucket to descend: the bucket rope was twisted about a spindle which, in turning, caused the gates to swing back. The æolipile, though it was no more than a toy, was the first steam engine known to history—and jet propelled at that!

Whatever the aim of this flourishing Alexandrian school of invention, it took no steps to spread knowledge of science and mechanics beyond its own walls. The coming of Christianity did not at first do anything to change the position. The early Christians set their hopes on the joys of the future life rather than on the possibilities of the present. Investigation into the facts of nature had the smell of devilry about it. Even the priests feared to dabble in it.

The great Roman Empire was breaking up. Relying on slave labor and enrichment by conquest, the Romans did not concern themselves with the everyday life of their subject peoples. Though they were splendid civil engineers, and built roads, bridges, and aqueducts the like of which were not seen again for many centuries, mechanical engineering—the making of power engines and machinery—was virtually unknown to them.

In the feudal form of society that followed, every village community provided its own necessities. The village had no need of large power-producing engines, nor of science. Its great need was manual skill and craftsmanship: not surprisingly, such skills flourished and developed, and with them the small hand tools which the craftsmen used.

Muscles, of horses, oxen, and men, were still the main sources of power. Milling was the only work for which some other source of power was commonly used. Domesday Book, compiled at the end of the eleventh century, lists about 5,000 mills in England, which would be sufficient for a total population of two or three million. These were all water mills: windmills were not introduced into this country until the following century, when the method was brought back from the East by returning Crusaders.

The village mill was owned by the lord of the manor, or else by the Church. The mill owner allowed no other mills to be built in competition with his. He charged every villager a milling fee, and must often have done very well out of his monopoly.

In the East, a discovery had been made which was destined to have the most profound consequences for the whole world. With this discovery begins the long and exciting story of the winning of power from heat, on which the development of our present day civilization chiefly depends.

In certain districts of the East, a natural substance called saltpetre lies on the surface of the soil. When primitive peoples lit their fires on the saltpetre deposits, and the embers burned down, carbon from the charcoal mixed with the saltpetre. At the touch of flame, the mixture sparkled and crackled. As human intelligence developed, men came to realize that they could make the sparkling mixture by intention, and delight everybody by producing its effect at will. Someone rolled up some of the mixture in dried leaves, and found that when he set fire to one end of the roll it jumped fizzing and crackling into the air.

In some such way, we may suppose, fireworks came into

being, and became a regular feature of Chinese life, and possibly of Indian life too. The jumping rolls became rockets. To the Chinese, these pretty toys were symbols of human feelings: they let them off to celebrate holiday occasions, weddings, and even funerals.

When Jenghiz Khan invaded China with his Mongolian armies, the symbols of feeling became weapons of defence. The Chinese loosed their rockets among the Khan's soldiers with terrifying effect. The Mongols were quick to take up the idea. They carried it with them on their conquering marches.

Some knowledge of the "incendiary mixture" had already reached the countries round the Mediterranean. In some unknown corner, an experimenter mixed a third substance, sulphur, with the saltpetre and carbon. The resulting combination, when wrapped tightly and kindled, did more than fizz and jump. It exploded with a thunderous noise and vivid flash.

Roger Bacon, the English monk, heard of this stuff when visiting Spain in the thirteenth century, and brought back to England the secret of its composition. Some say it was Bacon who first added the sulphur. At all events, he saw clearly enough that if the secret became known to everybody, much harm could be done. Partly for this reason, partly, no doubt, because he feared to be accused of treating with the devil, he recorded the formula in a secret code.

Within fifty years, the secret was out, and no time was lost in using it to do harm. Saltpetre, carbon and sulphur are the constituents of gunpowder.

It remained for someone to invent the gun. This was the achievement of another holy man, the German monk Berthold Schwarz. Realizing that the explosive force of the powder was of no use to anyone unless the direction of the explosion could be controlled, he solved the problem by making a strong metal tube, sealed at one end and open at the other. He rammed his powder tightly down against the sealed end, and confined it from the top with a heavy ball. When the mixture was fired, it had to expand somewhere, so it drove out its one movable barrier, the free-resting ball. The ball flew out to do its job of breaking down a castle gate, and the first practical heat engine, the cannon, was in being.

It was an "engine" because it produced effective power under control. Its purpose was purely destructive: it was incapable of giving a steady, continuous output of power. It operated in spasmodic bursts of great force, but it did the job for which it was intended more effectively than the thonged catapult or the muscle-powered battering ram.

What had happened meanwhile to the ingenious devices of Hero and his fellow scholars of Alexandria? Hero's works are a treasure house of inventive principles which did eventually play a significant part in the power story, but not until more than fifteen centuries had passed.

Much of the work the Alexandrians had done had to be done over again before it could be properly understood. Basic principles had to be rediscovered. Among the first to experiment systematically with water and heat and record his findings was an Italian, Giovanbattista della Porta. In 1600 he published a book that established certain very important facts. One was that when water turned to steam, the steam occupied very much more space than the water from which it had come. If it was not given the extra space, it pushed with terrific force to get it.

Della Porta made a metal container, to which the only outlet was an open pipe inserted through the top. The bottom end of the pipe was close to the floor of the container, so that when he poured water in, the pipe's lower opening was covered. Now he introduced steam from another vessel into the top of the container. The steam, pushing hard to give itself more room, thrust the water up through the pipe and out at the top in a strong jet.

It was easy to see that by reversing the process and condensing the steam back to water, the need for more space disappeared. But when della Porta started to examine this process, a new fact presented itself.

He filled a wine flask with steam, plunged the neck into a basin of cold water and waited for the steam to condense. As it condensed, the water in the basin began to rise up the neck of the flask until the flask was almost full of water. So long as the flask was held in this position, with its neck submerged, the water remained in it. What made the water rise, and why did it not drop back?

It was as if some force outside the flask was pushing against the surface of the water in the basin, and forcing it up into the flask. This was, in fact, what was happening. The atmosphere was pressing downwards on the water with a strength of 14 or 15 pounds to every square inch of its surface.

In such a way it was discovered that air, which seemed weightless, did in fact have weight. This piece of knowledge led to the next big step towards the winning of efficient power.

At first della Porta's principles were applied to unimportant purposes, like the working of fountains in noblemen's gardens. Early in the 1600's, a Frenchman named de Caus appeared at the court of James I of England, for whom he erected some of these ornamental steam fountains. The novelty interested King James' courtiers, one of whom, a Scotsman named Ramsey, found a way of using the principle to raise water from wells. His friend the Marquis of Worcester took the idea further. With the help of a Dutchman named Kaltoff, he made a "water-commanding engine" which, it was claimed, could throw a jet of water 40 feet in the air. Unfortunately he left no clear record of how it worked.

This is as much as is known of the early pedigree of the steam engine. As yet, the world was unaware of what was going on. Step by step, a new form of power was coming into being.

When Thomas Savery, an energetic and ingenious Devonshire man, constructed an engine worked by steam for pumping water from flooded mines, he claimed the invention for his own and took out a patent for it. Others said he had his idea from the Marquis of Worcester, which is probably true, for most inventions are the result of one man's improvement on another man's beginning.

Savery was a mining engineer, well acquainted with the difficulties which underground floods were causing in the

The Beginnings

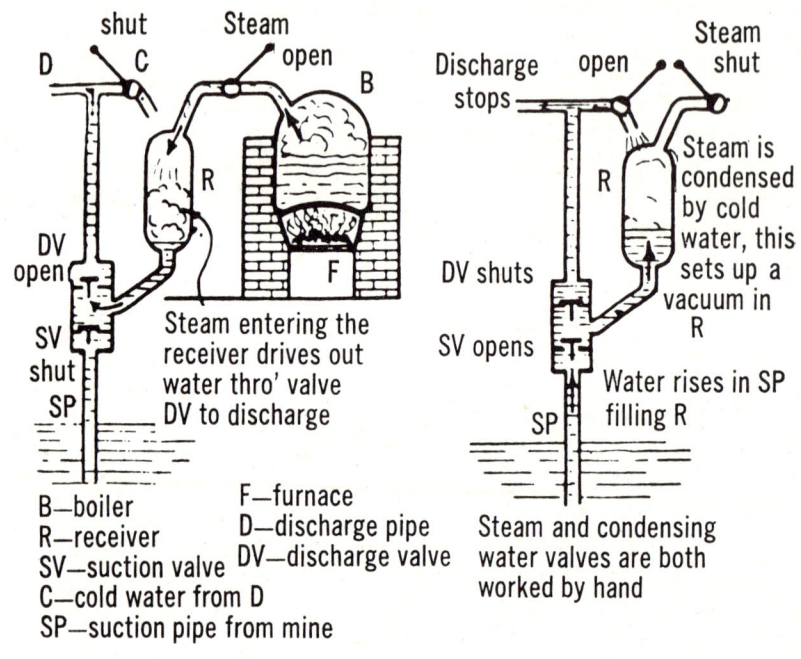

Diagram labels (left figure):
- D — shut
- C — Steam open
- B (boiler)
- DV open
- SV shut
- SP
- R (receiver)
- F (furnace)

Steam entering the receiver drives out water thro' valve DV to discharge

Diagram labels (right figure):
- Discharge stops
- open / Steam shut
- Steam is condensed by cold water, this sets up a vacuum in R
- DV shuts
- SV opens
- Water rises in SP filling R

B — boiler
R — receiver
SV — suction valve
C — cold water from D
SP — suction pipe from mine
F — furnace
D — discharge pipe
DV — discharge valve

Steam and condensing water valves are both worked by hand

I-2. *Savery's mine engine, 1698.*

coal and metal mines at this time. The situation throughout the country had become serious. The widespread use of wood for fuel was threatening rapidly to exhaust the timber supplies. At the very time when coal was urgently needed, many mines were coming to a standstill. The upper coal deposits had been worked out: as the pits went deeper in the search for new seams they passed through water-bearing strata from which the water seeped in faster than it could be baled out. The same problem faced the tin and copper mines of the west.

It was with this problem in mind that Savery built his pumping engine, which he called "The Miner's Friend." He was confident enough to establish a factory in the City of London, where for the first time in world history, orders were taken for the manufacture of steam engines. That was in

1702. But his engine, though workable, proved a feeble friend to the miner.

Savery put della Porta's two principles together. He made a pair of receiving vessels: while one was filling with steam, the steam in the other was being condensed by a spray of cold water directed against its outer surface. Water from below rose into the space left by the condensing steam, as it had risen in della Porta's flask. When the vessel was full, steam was admitted to it from a boiler, and pressed the water out through another pipe to drain away over the ground. A simple system of valves ensured that the water and steam passed through the right pipes at the right moment.

Apart from the valves, this engine had no moving parts. To bring the water up from below, it relied wholly upon suction, that is to say, on the pressure of the atmosphere on the surface of the water that was to be pumped out. Atmospheric pressure is strong enough to thrust water up about thirty feet of vertical pipe, and no more. There were difficulties about installing the engine at the bottom of the pit and leaving the main lifting pressure to the steam. Added to this, men did not know much about boiler-making in those days, and their boilers were apt to burst when steam pressure was high.

A new principle was needed. It had already been propounded, and Savery knew about it, but he did not believe it would work.

The great Dutch physicist and astronomer, Christian Huyghens, observing the power of a gunpowder explosion, conceived that it might be made to serve better purposes than destruction and death. About 1680, he set to work to build a gunpowder engine that could be put to a peaceful use.

The basic construction of his engine followed the general form of the cannon. The charge was to be fired in a tube, or cylinder, closed at the bottom end and open at the top. Instead of a cannon ball, he set a piston in the cylinder, which was free to move up and down but not to fly away and be lost.

The sudden expansion of the gases produced by an explosion was too abrupt a force to move a captive piston without risk of breakage. His plan was to use the explosion to drive the air out of the cylinder through leather valves, which would then close, leaving the cylinder full of hot gases. As the gases cooled, they contracted, like the condensing steam in della Porta's flask. The outer atmosphere, pressing on the cylinder from all sides, would move the only movable part, which was the piston. Thus the piston was thrust down to the bottom of the cylinder, where it remained until another gunpowder explosion filled the cylinder again and allowed the piston to return. The downward movement of the piston was the working stroke: the work was done by the pressure of the atmosphere.

By arranging a series of explosions and repeating the process, Huyghens hoped to have his piston moving up and down the cylinder in regular motion. Harness this motion to a machine, for instance the rods of a pump, and it could be made to do useful work.

In the event, gunpowder proved too uncertain a source of power for this form of mechanism. It was difficult to regulate the successive explosions evenly, and the leather valves quickly perished. The gunpowder pumping engine failed, but Huyghens had found the mechanical principle on which the whole subsequent development of power engines was based. So far as is known, he was the first to show how cylinder and piston might be used for the production of power. This was one idea that had not been foreshadowed by the Alexandrians. The greatest military victory has not had a profounder effect on human affairs.

A Frenchman with an inventive genius, Denis Papin, who came as a Huguenot refugee to England, was interested in Huyghens' engine, and tried to improve it by making more durable valves. The irregularity of the gunpowder explosions still defeated him: he looked round for a more controllable source of power, and picked on steam. To Papin in 1690 belongs the distinction of first putting steam and piston together; but he never got further than experiment and demonstration.

Savery, as we have seen, was working on other lines, which

in the long run proved unfruitful. Already when Savery was making his first engines, another Devon man was busy on a new type of construction. Thomas Newcomen, a blacksmith and ironmonger of Dartmouth, had come to hear of Papin's experiments with steam and piston. He, too, was well aware of the predicament of the metal mines, whose engineers he served. With his friend John Calley, a plumber, he set himself to turn Papin's theories to practical account. In 1712 he completed and erected the first successful steam piston engine known to the world—another momentous event.

It was a large, cumbrous mechanism, supported and housed by a tall brick building. At the bottom was the furnace, over which was placed a boiler with a round top, looking rather like a haystack or a beehive. To the top of the boiler was attached a large cylinder, open at the upper end, and communicating with the boiler below it by a valve through which steam was admitted. Above this, supported by the engine-house wall, was a large wooden beam that rocked see-saw fashion on a pivot. The piston was attached by a chain to the inner end of the beam, while the outer end was connected to the rods of a pump.

Following the principle used by Huyghens, and Papin after him, Newcomen did not use the expanding force of hot gases (steam, in this case) to do the effective work. For this, he relied on the weight of the atmosphere pressing in to fill the space left by the steam as it cooled and condensed. When steam was admitted to the cylinder, the weight of the pump rods, acting through the beam, pulled the piston up. A jet of cold water was sprayed into the steam-filled cylinder: the cold water condensed the steam, and down came the piston again, pulling up the pump rods in its turn. The various valves were opened and shut automatically as the beam rocked back and forth. So long as the furnace was stoked and steam admitted to the cylinder, the motion proceeded, slow but continuous. The pump rods alternately drew in water at the bottom of the intake pipe, and forced it up the outlet pipe, a more serviceable method than relying entirely on suction.

Newcomen's engine of 1712 was erected at Dudley Castle in Warwickshire, where it was employed to maintain a water supply within the building. Its undoubted success attracted the attention of the mine adventurers: it was not long before Newcomen pumping engines were at work in the main mining districts of the country. They saved many a pit from flooding.

The Dudley Castle engine had an output of about 5½ horse-power. From it, you may say, are legitimately descended the millions of steam piston engines that in one form

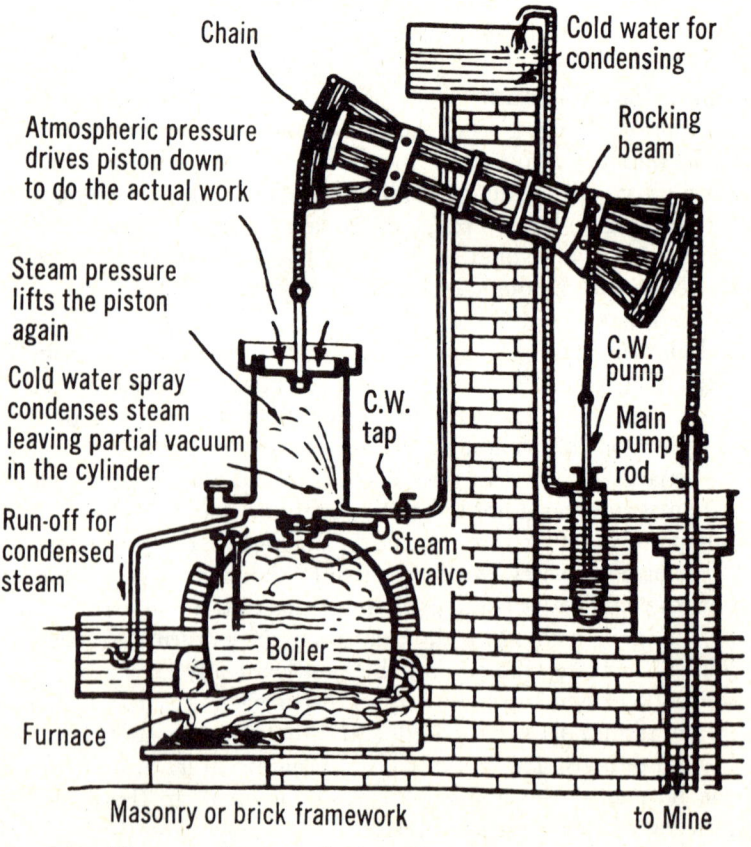

I-3. *Newcomen's atmospheric engine, 1712.*

or another have been building civilization ever since. No trace is left of it now, but it heralded the coming age of power.

Though atmospheric engines worked the pumps that kept the mines dry, the coals and ore were still wound to the surface by horse engines. The "cog and rung gin," and its more efficient successor the "whim gin," were given a winding motion by a team of horses tramping endlessly round a circle. Contemplating the two power sources side by side, many a mining engineer must have wondered whether there was not some means of setting steam to turn a wheel.

The nearest they came to it at this stage was to throw the water pumped up by the atmospheric engine on to a mill wheel, and so to provide rotary motion. The time was ripe for a more efficient installation than this.

In 1759, a young mathematical instrument maker named James Watt was working at his trade in the precincts of Glasgow University College when a friend came in to tell him of his latest inspiration. This was a plan to construct a "horseless carriage," driven forward by the power of steam. Watt set his diligent and subtle mind to examine this proposal, and found it unworkable; but he dearly loved a difficult problem, and seems to have realized that he must know more about steam and its working before he could come to any satisfactory conclusions. Hearing that the University's demonstration model of an atmospheric engine was in need of repair, he took it in charge.

Watt was 23 at the time. This was his first introduction to the subject of steam and what could be done with it.

Working on the Newcomen model, Watt was quick to discover its limitations. The fact that the cylinder became hot and cold by turns—hot when it was filling with steam, cold when the condensing jet was turned on—inevitably slowed down the movement, and spent fuel lavishly meanwhile. The first steam to enter the chilled cylinder condensed and was wasted: in the same way, the cold jet when first admitted was vaporized by the steam. The action had always to wait until the temperature of the cylinder had changed one way or the other.

Walking on Glasgow Green one Sunday morning in 1765. Watt hit on the solution to this problem. In the words of one of his friends, it "filled him with rapture." Judged by its eventual results, it proved to be one of the most significant ideas ever born in the mind of a man.

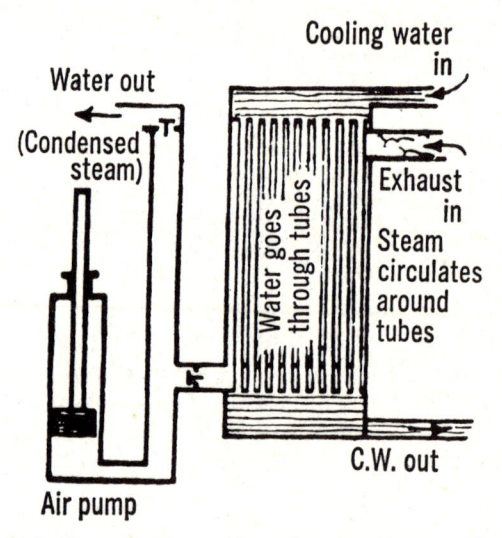

I-4. Watt's first experimental condenser, about 1766.

It was simple enough. Instead of condensing the steam inside the working cylinder, as Newcomen did, Watt proposed to attach a separate vessel to the cylinder, and condense the steam there. A vacuum was kept in the condensing vessel by means of an air-pump. When the cylinder was full of steam, a valve opened, and the steam rushed into the cold condenser to fill the vacuum. As soon as it entered, it was turned to water: the vacuum remained unfilled, and all the steam disappeared into it, until there was vacuum in the cylinder as well. The piston descended, and the working stroke was made. Steam was then let in for another stroke. Since the cylinder had never lost its heat, the piston rose immediately, without waste of time or fuel.

This was Watt's secret. His condenser was housed in a water tank to keep it perpetually cold, his cylinder was surrounded by a steam jacket to keep it perpetually hot. It was an improvement with revolutionary consequences.

Watt lost no time in experimenting, but it was not until 1769 that he was confident enough to take out a patent. Even then, for lack of skilled workmen, he failed to make a successful engine. Had not Matthew Boulton of Birmingham offered him a partnership in the making of steam engines to the new design, he would have given up his cherished ambition. Plagued by headaches, debt, and depression, he was more than once on the point of doing so.

He could not have been more fortunate in his new partner. Boulton had the confidence and spirit that Watt lacked. He was a man of energy, kindliness, and culture; moreover he had gathered at his Birmingham foundry some of the best mechanics in Europe. By 1776 the new "condensing engine," as it came to be called, was triumphantly at work.

Its appearance was timely in all respects, for by now many mines were going too deep even for Newcomen's atmospheric engines to keep them dry, while others found the fuel costs ruinous. The copper miners of Cornwall were at their wits' end, and here it was that Boulton and Watt first began to win profitable returns.

It was the far-sighted Boulton who first put into Watt's mind the idea of setting the condensing engine to turn a wheel. Watt would have been happy to have rested in modest security on the success of his engine as an improved steam pump. Boulton could see that with factories going up on a scale never dreamed of a century before, the demand for wheel engines would grow faster than the need for new pumps. Manufacturers were waiting for such an invention. The mere prospect, said Boulton, had sent cities like London, Birmingham and Manchester "steam mill mad."

Watt's first "rotative" was completed for John Wilkinson, the ironfounder, in 1783. Then the partners erected a steam-driven flour mill on Thames side in Southwark, the fame of which went far. Orders for rotatives began coming in from all quarters.

From now on, industry was no longer tied by water wheels to rivers, or brought to a standstill by frost and drought. James Watt had made a power unit that could be applied to any kind of standing machinery in almost any place. After that, factory production began in earnest.

This was the triumphant outcome of that masterly inspiration on Glasgow Green of twenty years earlier. As a result of it, the way of the world was changed.

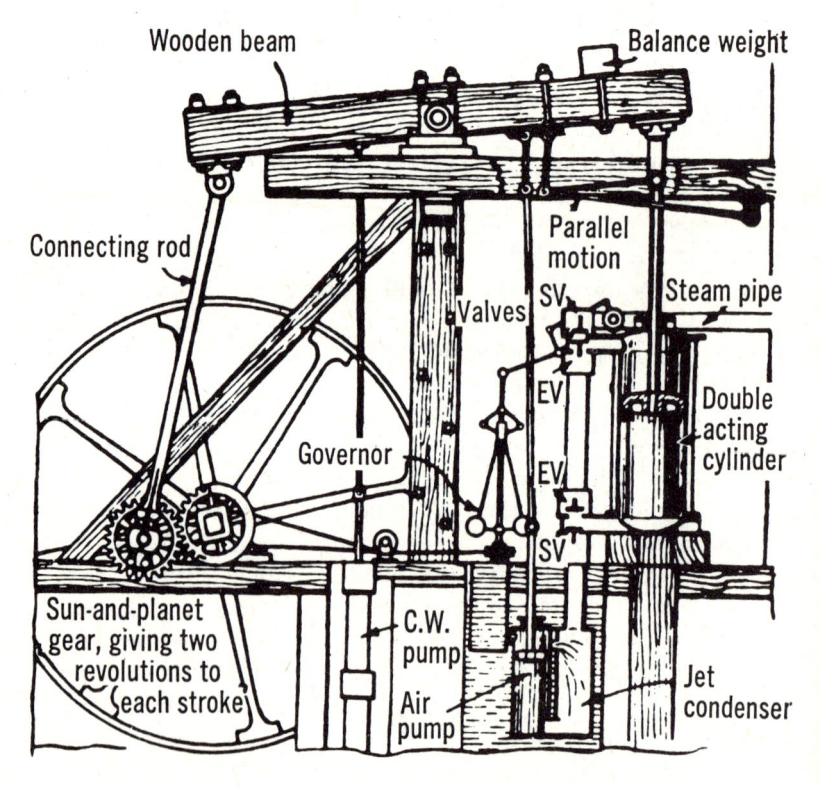

I-5. Watt's rotative engine, 1783. This was the first engine to make successful use of artificially-produced power for turning a wheel. After this date, it was no longer necessary to rely on water and wind for driving machinery. Watt's rotative engine, which can be studied in the London Science Museum, marks the greatest turning point in the story of power.

It has been noted that the science of engineering has owed much of its progress to a multitude of practical men. The names of most of these individuals are buried in obscurity and with a few exceptions those of ancient times are completely unknown. In engineering, as in mathematics and physics, Archimedes stands preeminent. He was born in Syracuse about 287 B.C. and was slain by a Roman soldier during the siege of the city in 212 B.C. His is generally considered to have been the greatest scientific mind of ancient times. Among his contributions to engineering were a statement of the principle of the lever ("Give me a place to stand and I will move the earth"); the Archimedean screw for lifting water; the compound pulley for raising weights; and numerous military devices, including catapults and burning glasses, which terrified the invaders who finally overran the city.

The names of other engineers are scattered through ancient and medieval records—Imhotep, the Egyptian master of stone construction; Eupalinus of Megara, who, according to Hippocrates, constructed a tunnel under a mountain on the island of Samos; Hippodamus, one of the first city planners, who redesigned the layout of the port of Piraeus; Hero, whose writings are described in Miles Tomalin's article on page 14; Appius Claudius, who gave his name to the Appian Way and the Appian Aqueduct; and Frontinus (the hydraulic engineer who has been quoted on page 2).

In engineering, as in basic science and the arts, the Renaissance saw the appearance of a group of protean geniuses. Among them were Michelangelo, whose bronze castings were a landmark in metal working and whose plans for the beautification of Rome, including the great base of the central dome of St. Peter's, entitle him to be classified as one of the great master builders of all time; Brunelleschi, whose massive circular dome for the cathedral of Florence, constructed with two shells and an octagonal vault, is generally considered his masterpiece; and Palladio, whose trussed bridges compare favorably with much later work. Even among this distinguished group, the star of Leonardo da Vinci shines brightly. Bern Dibner, an electrical engineer and American historian of science, here describes one aspect of his multitudinous activities.

LEONARDO DA VINCI: MILITARY ENGINEER

BERN DIBNER

IF, ON SOME SUNNY Saturday morning, Leonardo da Vinci were invited to appear at one of the regular meetings of the joint chiefs of staff in Washington, much of the conversation would, of course, be strange to him, but generally he would follow the comments and reports not as a stranger. Recommendation for a barrage to be laid down here, combined operations there, tank assault following softening from the air, chemical warfare at this point, the success of submarine attack—all this would pass through his mind, and the important points would go down into his ever-present notebook, with a sketch where possible. Should the discussion turn to the big elements, big guns, big areas, great numbers, he might even drop a well-considered word. Discussions of electronics alone would leave him wondering.

This man not only lived and thought in terms many centuries ahead of his times, but what is more, he left a record of his plans and thoughts. These lay buried for some four hundred years, unappreciated and unpublished, because the affairs of men had not yet caught up with his advanced thinking. Now we wonder how one mind could have held it all.

Leonardo was a specialist in many fields of science. In some, such as perspective, anatomy, mechanics, botany, and hydraulics he is considered among the very best. Although Leonardo hated war, as all rational men hate it, yet there seemed no escape from its "bestial madness," to quote him.

Leonardo held three appointments as military engineer. The first was by Ludovico Sforza, Duke of Milan, to whose court Leonardo came in about 1483. The second was by Cesare Borgia, at the height of his military adventure in 1502. The third was at the invitation of the grand monarch,

Francis I, in 1516, when Leonardo was a spent old man. This was his last mortal triumph, and it was short-lived.

What prompted the coming of Leonardo to the Milanese court of Ludovico is not definite, there being reason to believe that he came as a musician or as a painter. Most probably he came as a military engineer, to judge from the evidence that the existence of a copy of a letter offers. It reads:

> I can construct bridges very light and strong, and capable of easy transportation, and with them pursue or on occasion flee from the enemy, and still others safe and capable of resisting fire and attack, and easy and convenient to place and remove; and methods of burning and destroying those of the enemy.
>
> I know how, in a place under siege, to remove the water from the moats and make infinite bridges, trellis work, ladders and other instruments suitable to the said purposes.
>
> Also, if on account of the height of the ditches, or of the strength of the position and the situation, it is impossible in the siege to make use of bombardment, I have means of destroying every fortress or other fortification if it is not built on stone.
>
> I have also means of making cannon easy and convenient to carry, and with them throw out stones similar to a tempest; and with the smoke from them cause great fear to the enemy, to his grave damage and confusion.
>
> And if it should happen at sea, I have the means of constructing many instruments capable of offense and defense, and vessels which will offer resistance to the attack of the largest cannon, powder and fumes.
>
> Also, I have means by tunnels and secret and tortuous passages, made without any noise, to reach a certain and designated point; even if it is necessary to pass under ditches or some river.
>
> Also, I will make covered wagons, secure and indestructible, which entering with their artillery among the enemy will break up the largest body of armed men. And behind these can follow infantry unharmed and without any opposition.
>
> Also, if the necessity occurs, I will make cannon, mortars and field-pieces of beautiful and useful shapes, different from those in common use.

Where cannon cannot be used, I will contrive mangonels, dart throwers and machines for throwing fire, and other instruments of admirable efficiency and not in common use; and in short, according as the case may be, I will contrive various and infinite apparatus for offense and defense.

In times of peace I believe that I can give satisfaction equal to any other in architecture, in designing public and private edifices, and in conducting water from one place to another.

Also, I can undertake sculpture in marble, in bronze or in terra cotta; similarly in painting, that which it is possible to do I can do as well as any other, whoever it may be.

Furthermore, it will be possible to start work on the bronze horse, which will be to the immortal glory and eternal honor of the happy memory of your father, My Lord, and of the illustrious House of Sforza.

And if to anyone the above mentioned things seem impossible or impracticable, I offer myself in readiness to make a trial of them in your park or in such place as may please your Excellency; to whom, as humbly as I possibly can, I commend myself.

After the dissolution of the reign of the house of Sforza in Milan in the very last days of the 1400's, Leonardo went to Venice, there noting elements in naval ship construction and of naval warfare. He was in the company of Luca Paccioli, the mathematician, for whose book on the mean proportion Leonardo designed the mathematical figures. He then drifted back to Florence, and in 1502 took service with Cesare Borgia as his military engineer. Service under Cesare Borgia took Leonardo to Faenza, Urbino, Imola, Pavia, and other cities during the campaign of the Romagna. As part of his work, Leonardo prepared topographical maps of cities and whole districts of central Italy. These are remarkable in their fidelity and detail and excellent for their intelligence data. In July 1503, Leonardo was with the troops of the city of Florence in a siege before Pisa, during another one of a long series of wars between these two cities, both located on the same river, the Arno.

The third military appointment was an honorary one, from

the young and powerful monarch of France, Francis I, appointing Leonardo his *"ingénieur, architecte, méchanicien."* This admirer of all of Leonardo's talents had hoped not only to bring him to retirement and peace in a hospitable France, but to have all the master's work brought there as well. Moving the wall on which the Last Supper was painted proved an impossible task, so it remained in Milan. Many other paintings traveled with the frail old man over the Alpine passes, and these have remained in France ever since. Leonardo was commissioned engineer-architect to the royal house, and for three years, 1516 to 1519, Leonardo was made comfortable. Here at St. Cloux, in Amboise, he worked on several canalization schemes, including that of the Romarantin, and as advisor to the expanding construction program of the vivacious Francis. The sunset years were few, for even in the welcome peace of his retirement, the many active and creative years had left Leonardo old, his right arm useless. Here in Amboise, an ever friendly France took his tired frame to her.

Leonardo's military engineering works divide themselves into three general categories—engines for offense, structures for defense and general equipment for construction, excavation, mining, ordnance and for the services of supply.

In devising engines that would give strong offensive action, greater fury, firepower and speed, Leonardo looked to increased calibre of cannon, better firing means and more rapid fire from multiple guns. He reviewed the ancient writers and sketched their descriptions of tactics and equipment. One device that catches the imagination by the combination of its cleverness and horror is the arrangement of whirling scythes shown in several sketches in the Paris manuscripts. Horse-drawn chariots have their wheels geared to whirling scythes that dash through a field of men, mowing them down.

On the same paper containing the scythed chariots, one finds what has often been called a tank. It is really intended as an armored, self-propelled vehicle, lacking the basic feature of the modern tank, its tread. However, for its time, a highly ingenious device is shown. The left view shows the top removed. The four drive-wheels are rotated by two cranks, each crank turns two lantern pinions which engage the gear-points on the wheel faces. The whole vehicle is topped by a protective cover of planks, with a cap on top. Two horizontal firing slots are provided, so that men inside may fire as the "tank" moves, like a kind of land-ship reminiscent of the *Merrimac*.

A test of Leonardo's inclination to use mechanical power instead of muscular power, even in an age of cheap and abundant labor, is shown by his design for the canal cutting engine. In Leonardo's time, as in our own, canals were used to divert rivers, flood valleys, drown out armies, destroy communications and supplies. Leonardo, in his notes, has several recommendations on methods of using canals for the diversion of rivers as an offensive weapon.

For assault troops, Leonardo designed light portable bridges that were to be moved in elements and then assembled at the river edge or at the moat. Similarly hinged, swiveled and cantilever bridges, quickly constructed and easily swung across the river or canal, are sketched in his notebooks.

Fortifications are massive, of polygonal design, with high lookout tower and bastioned angles. Leonardo devised a means for dislodging scaling ladders leaned by an enemy against the top parapet. Vertical pivoted bars are pulled away from the inner wall by defense troops. In so doing, the tops of the bars push inwardly and thereby force out a hidden horizontal bar set flush with the escarpment. This bar moves out, forcing the ladders away and pitching them beyond their footing.

As indicated in his letter to the Duke of Milan, Leonardo had advanced notions on mining and countermining. This was one way of breeching a moat-protected fort without direct assault. Against such mining, he recommends detection of the sound of the sappers at work by placing a drum above the area where the enemy is suspected of mining. Upon the drum are placed a pair of dice, "and when you are near the place where the mining is, the dice will jump up a little on the drum, through the blow given underground in digging out the earth."

Only a few of the great number of engines and mechanisms proposed by Leonardo for military use are here described. This is equally true of the other groups of military equipment.

The imaginative Leonardo perceived a method of refining the manufacturing methods of boring by reducing the tolerances in order to increase the fire-power and range. Advice to the founders of bronze cannon is contained in manuscript M 54v:

> That part of the bronze is most compressed within its mould which is most liquid. And that is most liquid which is hottest, and that is hottest which comes first out of the furnace. One ought therefore always to make first in the casting that part of the cannon which has to receive the powder before that which has to contain the muzzle.

In artillery, Leonardo's drawings and notes indicate several classes of equipment: mortars, cannon, great slings, catapults and crossbows, field guns, and several types of machine guns. Some pieces of artillery were conceived of such great dimensions as to be impossible to transport over the terrain and roads of his day.

Many of Leonardo's sketches show slings and catapults, using the energy stored in bent and twisted wooden arms, in metal springs and in the torque of twisted ropes by the use of worms and gears, or racks and pinions. Many of these designs appear in older sources, for Leonardo transferred to his notebooks any that were of interest to him.

It was one hundred and fifty years from that fateful day at Crècy, when two cannon, attended by twelve English cannoneers, first fired a shot propelled by gunpowder at an enemy in battle. Yet a casual examination of some of Leonardo's sketches would indicate that his concerns were those of an artillerist of the mid-1800's. In some respects he was ahead of even those days. He was, for instance, concerned with good shrapnel design, with breech instead of muzzle loading, with ease and speed of fire, with multiple fire and with advanced gun construction methods. The shot is contained in a leather cover, and a fuse ignited at firing causes the shell to burst in the air.

The use of a complete prefabricated cartridge was to Leonardo as obvious an expedient as was the advantage of breech loading.

An unusual piece of artillery was designed and termed "architonitro" by Leonardo. This consists of a cannon that depended on the sudden generation of steam to drive the shot out of the barrel. The breech was built into a basket-like brazier, containing the burning coals. With the shot rammed back, and the breech section sufficiently heated, a small amount of water is injected into what would normally be the powder chamber.

> And when consequently the water has fallen out it will descend into the heated part of the machine, and there it will instantly become changed into so much steam that it will seem marvellous, and especially when one sees its fury and hears its roar. This machine has driven a ball weighing one talent six stadia.

This would indicate that such a gun had actually been made. Evidently the rate of fire was a lesser consideration. Putting steam's expansive power to work was novel indeed, in Leonardo's time.

Living his life inland, as at Florence and Milan, and dying inland (at St. Cloux in France in 1519) Leonardo yet shared with his contemporaries a strong interest in things pertaining to water and the sea. So valued were his notes and drawings on the motion of water and the flow of rivers, that these were gathered and published more than a century ago. One of his ideas—how a man can remain under water—was used by several writers of military texts in the 1500's and 1600's, and they credited Leonardo with the invention. On the other hand, an even more effective idea was withheld by Leonardo from his own notebook, for the reason he gives:

> How by an appliance many are able to remain for some time under water. How and why I do not describe my

method of remaining under water for as long a time as I can remain without food; and this I do not publish or divulge on account of the evil nature of men, who would practice assassinations at the bottom of the sea by breaking the ships in their lowest parts and sinking them together with the crews who are in them.

For the annihilation of enemy shipping Leonardo proposed the use of Greek fire, a weapon first used in the seventh century. This was a combination of sulfur and quicklime, which on touching water was ignited. Another proposal was to

> Throw among the enemy ships with small catapults, chalk, pulverized arsenic and verdigris. All who inhale this powder will be asphyxiated by breathing it, but be careful that the wind be such as not to blow back the fumes, or else cover your nose and mouth with a moist cloth so that the powder fumes cannot penetrate.

Leonardo's notes on the flight of birds and of his proposed flying machine are scattered in many of his notebooks and on odd pages of notes, for the thought of mechanical flight was one of his most persistent:

> I have divided the *Treatise on Birds* into four books; of which the first treats of their flight by beating their wings; the second of flight without beating their wings and with the help of the wind; the third of flight in general, such as that of birds, bats, fishes, animals and insects; the last of the mechanics of this movement.

As the function of each component part of a bird is broken down to show its function in flight, so the kinds of flight are grouped into straight, curved, spiral, circular, falling and the complex combination of these.

In reading Leonardo's notes on the construction of the various versions of his flying machine, and those on the flight of birds, it is difficult to tell to which drawing he refers. His machine, an ornithopter, was intended to closely follow the successful flight of birds he saw constantly about him. Like his studies of different kinds of flight, so his keen sight and patient analysis broke each flight cycle into its elements of motion of beak, head, body, wings, tail and feet. The same detailed breakdown of motion that each one of his airplane elements was to make was similarly studied under conditions of ascent, soaring, in turbulent air and in descent.

The powerful chest muscles that Leonardo knew to be the basis of a bird's or bat's ability to fly must have an equivalent in man if he also were to be capable of flying with the aid of artificial wings. Did man have the required strength to equal that of the bird? Leonardo believed that if he could put man's powerful leg muscles to operating the wings, a sufficient source of power would be made available. The bird's muscles, he believed, were not only to sustain flight but they also had a triple factor of safety, in order to acquire a speed of flight necessary to attack its prey. Man, not requiring this, surely had sufficient power for only sustained flight. Leonardo, therefore, designed his wings to be operated by the leg muscles rather than depending on the chest muscles alone.

That actual flight was in Leonardo's mind is reflected by his notes on safety. There is the parachute, the first known reference to such a device. "If a man have a tent roof of which the pores have all been filled, and it be twelve braccia square and twelve in depth, he will be able to let himself drop down from any great height, without suffering any injury." Then the fear of falling from land into water is covered by a note recommending a double chain of leather bags, presumably air filled, to be "tied underneath, you so manage that these are what first strike the ground."

Soaring upon a wind is different from flight, because the source of the energy required to counter gravity has changed.

> But when the bird finds itself within the wind, it can sustain itself upon it without beating its wings, because the function which the wing performs against the air when the air is motionless is the same as that of the air moved against the wings when these are without motion.

Also, as written in another notebook, "it takes as much

to move the air against the immovable object as to move the object against the immovable air."

With the behavior and properties of air clearly in his mind, Leonardo's imagination extends itself to conceiving and designing a helicopter.

I find that if this machine made with a helix is well-constructed, that is to say made out of linen of which the pores are closed with starch and it is turned with great speed, the said helix is able to make its spiral in the air, and it will rise high.

The helix was to be eight braccia in radius, and the linen was to be stretched on stout cane. He recommends that a small pasteboard model, if dropped, will demonstrate the reverse action of helix and the air, in that in falling the helix will turn.

Many are the schemes that Leonardo devised to make the flying machine work. In some, the flyer rests prone; in others, upright, the latter facilitating take-off by leaving the feet free to run for the start. For strength and pliability, structural members are to be of laminated hemlock or fir, springs of laminated steel, or of cow's horn, hinges of leather. Wings were to be made of netting with shutters twenty braccia in length and breadth to permit wing-lift with reduced resistance. These were to open on the upbeat and shut on the downbeat. Having designed and made the wing, its effectiveness must be tested. Therefore, a testing block is rigged up to determine the difference in behavior between upbeat and downbeat.

One of the large halls in the Palace of Discovery at the world exposition held in Paris in 1937 honored the great physicists of all time. Large operating models demonstrated some basic mechanical principles and held the attention of the visiting hundred thousands. Upon the walls were framed the portraits of six outstanding men in the mechanical sciences—Aristotle, Archimedes, Leonardo, Galileo, Newton, Lorenz—a noble company. This is one more token among many of Leonardo's work in the field of mechanics becoming recognized as of a highly creative order.

II. The Age of the Machine

A. Power

B. Materials and Methods

II. The Age of the Machine

A. Power

The harnessing of power has been perhaps the most important single function of the engineer and the most revolutionary in its effects. It has been indicated that the availability of power—far more than ethical or religious considerations—resulted in the abolition of slavery. Power has helped us increase manyfold our production of food, clothing, shelter, and all the other physical needs of man. On it depends the standard of living of the Western World. The thesis that the height to which a civilization has risen is measurable by the amount of power it produces can be strongly defended on a practical —though obviously not on an ethical or intellectual—level. The first great revolutionary advance in its use stems largely from the invention of the "mathematical instrument maker," James Watt.

By the third decade of the eighteenth century, the Newcomen engine had spread from England to the Continent, had largely replaced draught animals wherever its use was possible, and had greatly increased mining and manufacturing production. Interestingly, one of its commonest uses was to pump water to supply a head for water wheels. Slow and inefficient though it was, it had no equal as a prime mover. Yet it was inadequate to meet the needs of the expanding economy. The industrial revolution, begun with the three nonhuman sources of power which have previously been mentioned, was gathering headway. Despite wars and political upheavals, the most important of which was the French Revolution, international trade was increasing, spurred by improvements in transportation and the availability of capital. Colonial exploitation was a factor in increasing the demand for goods and the rewards of industry. A new intellectual ferment was abroad and a new scientific curiosity was scrutinizing natural law—with profound repercussions on engineering. Greater industrial efficiency was a factor in the doubling of the world population between 1750 and 1850, and in turn resulted in more pressing demands for food, shelter, and cloth. James Watt, born in 1736 in Greenock, Scotland, was one of those fortunate individuals for whom the time was ripe. His father was a man of many trades and few means. The boy was sickly, studious, and unhappy. Several early apprenticeships ended in failure. Then, through the good offices of Dr. Dick, professor of philosophy at the University of Glasgow, young Watt obtained a position as mathematical instrument maker to the University. It was a chance which was to have profound effects on the world's economic development. The experiments by which Watt finally contrived his new machine for the development of power are here lucidly described by the author of a number of books on science and engineering (including The Great Engineers), who has been a Group Captain in the Royal Air Force and Deputy Director of Educational Services for the Air Ministry in Great Britain.

JAMES WATT AND THE HISTORY OF STEAM POWER

IVOR B. HART

AMONG THE POSSESSIONS of the natural philosophy class of the University of Glasgow was a working model of the Newcomen engine. It was not a good model, and apparently had never really properly served its purpose. The professor at that time was Dr. John Anderson, whose only scientific claim to fame lies in the fact that he was instrumental in arranging for Watt to handle and study the Newcomen engine model. The model had been sent in 1759 to a famous London instrument maker, Jonathan Sisson, for repair. However, when it was returned it was still unsatisfactory, and as a piece of demonstration apparatus was little more than useless.

During the college session of 1763-1764, Professor Anderson asked Watt to try his hand and skill at the task of its overhaul. He did so, he tells us, "as a mere mechanician, but then began to study it seriously." It was intended to be a scale model, but its boiler was incapable of providing enough steam to work the engine satisfactorily. The boiler had a diameter of about nine inches, the steam cylinder was two inches in diameter, and the stroke of the piston was about six inches. The whole, complete with cistern, model beam, and pump, was mounted on a simple frame and stand.

Watt tells us that "by blowing the fire it was made to take a few strokes," and then it gave out. Clearly, here was something for study, and with characteristic clarity he resolved the matter into two primary problems. One was simple enough. It was merely that of understanding the general principles of the Newcomen atmospheric engine, with its utilization of the power of atmospheric pressure against evacuated space produced by the condensation of steam within the cylinder on the other side of the piston—brought about, it will be recalled, by the introduction of a spray of cooling water inside the cylinder at the appropriate moment at the end of the upward or outward stroke of the piston.

The other primary matter followed. Having mastered the general principles of the engine, he quickly realized that the failure of the model arose from waste—that, although energy was available from the steam of the boiler, a considerable quantity of it—three-quarters, he estimated—was lost through one cause and another, so as to render the design of little use.

In asking himself what were the elements in the theory and working of the Newcomen engine that involved waste of steam power, Watt decided that the cylinder itself held most of the secret. When steam is admitted from the boiler, it has to enter as hot as possible and to stay as hot as possible. But he soon saw that some of the steam power was wasted through being used in supplying heat to the walls of the cylinder itself. To reduce this element of waste to a minimum, Watt decided that the material of the cylinder must be either a good conductor of heat, with its walls kept as thin as possible, in which case it needed very little heat to bring its temperature up almost instantaneously to that of the steam itself; or alternatively, it must be a thick-walled bad conductor like wood, in which case it would be absorbing the heat from the steam too slowly to matter, always provided that the speed of working was sufficient, so that it would not have time to interfere with the proper purpose and function of the steam.

Actually, for many reasons, the use of wood cylinders was bad and impracticable; and so far as the good conducting metallic cylinders were concerned, brass, the original material of the Savery engine, was undoubtedly good, but too costly. Cast iron had to be accepted as a compromise between cost and good conductivity, provided the size in relation to the length of the stroke and also the rate of working was economically calculated. But there was another important consideration. Bearing in mind that the steam is admitted from the boiler to fill the cylinder during the upstroke of the piston, i.e., when the weight of the pump rod is effective in swinging the crossbeam so as to draw the piston of the engine upward, the next step is to introduce a spray of cold water into the base of the cylinder to produce a vacuum. Here again was another source of waste. For some of the "cold" was being used not on the hot steam but on the hot walls of the cylinder and piston, and just that much of it was wasted through not concentrating on the sole job of condensing the steam. Moreover, for a maximum of effect and efficiency, *all* the steam in the cylinder must be condensed to produce a complete vacuum. In practice it never was. Some of it remained uncondensed, and thereby set up a certain amount of back pressure, giving that much less force on the piston for its downward stroke; and therefore just that much less power for the pumping up of the water, if that (as it usually then was) was the function of the engine. So Watt was prompted to the scientific investigation of the steam itself, to see how the knowledge of its properties would help him to surmount these defects and difficulties.

His experiments on the power and properties of steam are worth considering in some detail. As a preliminary, he first considered the problem of the induction water and its task of condensing the steam in the cylinder. Watt expressed this point in the following terms:

> It was also found that all attempts to produce a better exhaustion by throwing in more injection, caused a disproportionate waste of steam. On reflection, the cause of this seemed to be the boiling of water in vacuo at low heats, a discovery lately made by Doctor Cullen and some other philosophers, below 100°, as I was then informed, and consequently at greater heats, the water in the cylinder would produce a steam which would, in part, resist the pressure of the atmosphere.

Watt's next step was to discover for himself by experiment the relationship between the volume of a given quantity of water at the temperature of its boiling point, i.e., at the moment before it starts to boil, and the volume of the steam which this same quantity of water produces when it has all boiled away, still at the temperature of its boiling point. The result came to one cubic inch of water producing about 1,800 times its own volume of steam.

Watt was impressed by the enormous heating power of steam, and realized that at the temperature of boiling water the steam during its process of condensation was giving up or supplying a considerable quantity of heat to many times its own weight of water.

In the notes by Watt on Robison's *Dissertation on Steam-engines*, he describes his experiment thus:

> A boiler was constructed which showed, by inspection, the quantity of water evaporated in any given time, and thereby ascertained the quantity of steam used in every stroke of the engine, which I found to be several times the full of the cylinder. Astonished at the quantity of water required for the injection, and the great heat it had acquired from the small quantity of water in the form of steam which had been used in filling the cylinder, and thinking I had made some mistake, the following experiment was tried: A glass tube was bent at right-angles; one end was inserted horizontally into the spout of a tea kettle and the other part was immersed perpendicularly

in well-water contained in a cylindrical glass vessel, and steam was made to pass through it until it ceased to be condensed and the water in the glass vessel was become nearly boiling hot. The water in the glass vessel was then found to have gained an addition of about one-sixth part from the condensed steam. Consequently water converted into steam can heat about six times its own weight of well-water to 212°, or till it can condense no more steam.

As a rough experiment (which Watt subsequently repeated with much greater accuracy) this was a good pioneer result. Remembering that it takes 100 heat units to heat one gram of water from 0° C. to 100° C., this means that, if we take the normal temperature of the water in the cylindrical vessel as, say 11° C. (corresponding to about 52° F., which it probably was), the amount of heat required to bring one gram of this to the boiling point of 100° C. would be 89 heat units; and six times this would give 534 heat units for the latent heat of steam; and the accurate figure of today is 537 heat units. As Watt said, "Being struck with this remarkable fact and not understanding the reason of it, I mentioned it to my friend Doctor Black, who then explained to me his doctrine of latent heat, which he had taught for some time before this period [summer, 1764]."

Watt himself has summed up for us the researches to which his original handling of the famous Newcomen engine model led him in a very thorough and complete introduction to the next great step toward the invention of his steam engine:

1] The capacities for heat of iron, copper, and of some sorts of wood, as compared with water.
2] The bulk of steam compared with water.
3] The quantity of water evaporated in a certain boiler by a pound of coal.
4] The elasticities (i.e., pressures) of steam at various temperatures greater than that of boiling water, and an approximation to the law which it follows at other temperatures.
5] How much water in the form of steam was required

every stroke by a small Newcomen engine with a wooden cylinder six inches in diameter and twelve inches stroke.

6] The quantity of cold water required in every stroke to condense the steam in that cylinder, so as to give it a working power of about seven pounds on the square inch.

In the light of all this research, what was the answer to the problem of the waste in the Newcomen engine of as much as three-quarters of the steam power provided from the boiler? Watt could now see that this was something considerably more than mere defects of detail: it was fundamental to the whole conception and design of the engine. Things were being expected from the cylinder which were incompatible with efficiency. As we have seen, when the cold spray was brought in to condense the steam to produce a vacuum, some of it was, instead, having to be used to cool the cylinder; and when, in the next phase of its working, more steam was admitted for the next upward stroke of the piston, some of the steam was inevitably wasted through having to come in contact with the new cold walls of the cylinder. The steam, in other words, that ought to be used solely on the piston was also being partly employed in heating the walls of the cylinder back again to the temperature of boiling water.

Watt was faced with a seeming paradox. On the one hand, it would seem that the injection jet had to be as small as possible in order to keep down the wasteful "cooling" effect on the cylinder, and to confine it to its proper task of condensing the steam. But if this was pushed too far, the danger was that, instead of the cold water of the jet condensing the steam, the hot steam might so quickly heat the small spray of the jet as perhaps to vaporize it. That was what gave rise to back pressure on the wrong side of the piston. On the other hand, we have seen that water boils at below 212° F. when the pressure is below normal. Now, as soon as the injection cock is turned on, the spray begins to condense the steam. This means that as the vacuum tends to become more complete the pressure falls rapidly, and therefore the temperature at which the injection water boils also falls rapidly to a temperature as low as 100° F. If, therefore, we are to avoid this wasted energy of back pressure we must introduce only such an amount of cold injection water that the steam will not have time to heat up to 100° F. In other words, on this account the injection jet must be made as large as possible. Here, then, was the paradox. For one reason the jet must be kept small; for another reason it must be kept large. Now, the way of the engineer had hitherto been to compromise, and therefore to lose something "both ways." Watt was quite clear about the problem. How could he, at one and the same time, see to it that the cylinder could, ideally, be kept always at the temperature of boiling water, i.e., 212° F., to prevent steam in contact with it from being prematurely wasted by condensation, and yet also be kept at a temperature of less than 100° F., to prevent the injected water from vaporizing and producing the evil of "back pressures"?

Being clear about the problem was one thing; being clear about its solution was a quite different matter. We can well imagine how the problem must have teased and tantalized him. Before long it must have haunted his every waking moment. Then the answer came to him. Since the solution required that two separate and very different temperatures be maintained simultaneously, there must inevitably be needed two separate vessels for the purpose—the cylinder always kept hot at 212° F., and a completely separate cool vessel in which the steam, having been transferred from the hot cylinder for the purpose, could be separately condensed. Years later, Watt gave a personal account of this great moment of discovery to a Glasgow engineer named Robert Hart, who recorded it as follows:

> It was in the Green of Glasgow, I had gone to take a walk on a fine Sabbath afternoon. I had entered the Green by the gate at the foot of Charlotte Street—had passed to the old washing-house. I was thinking upon the engine at the time, and had gone so far as the Herd's house when

the idea came into my mind that, as steam is an elastic body, it would rush into a vacuum, and if communication were made between the cylinder and an exhausted vessel, it would rush into it, and might there be condensed without cooling the cylinder. I then saw that I must get quit of the condensed steam and injection water, if I used a jet as in Newcomen's engine. Two ways of doing this occurred to me: first, the water might be run off by a descending pipe, if an offlet could be got at the depth of 35 or 36 feet, and any air might be extracted by a small pump: the second was to make the pump large enough to extract both water and air. . . . I had not walked further than the Golf House when the whole thing was arranged in my mind.

We can picture him hurrying home, impatient of the fact that it was the Sabbath, but bowing of necessity to the convention involved. But the next day he lost no time in proceeding to test his ideas by means of a model—a thing of tinplate and solder, with a sewing thimble taken, no doubt, from his wife's workbasket, and bearing all the signs of crude and hasty work. It was a historic model, nevertheless. Briefly, Watt added to the engine a new constituent —an empty vessel, separate from the cylinder, into which steam should be allowed to pass from it and be condensed by the application of cold water either outside it or as a jet. Further, in order to preserve the vacuum in the condenser, Watt proposed to add an air pump whose function it was to pump from the condenser both the condensed steam and water and also the air that would otherwise accumulate from leakage or from being in contact with the steam or the injection water. Thirdly, as the cylinder was no longer to be used in the role of condenser, there was no longer any need to chill it. Watt proposed to keep it hot by surrounding it with nonconducting material, suggesting for this purpose a steam jacket between the cylinder and the outside casing. Lastly, and again with the same purpose, Watt proposed covering in the top of the cylinder, which up to now had always been open to the atmosphere, and he arranged that the piston rod should pass out through a steamtight stuffing box.

Consider how far we have traversed from the earliest days of engine design by various workers who brought something of a scientific approach to the task. First we have Denis Papin, who concentrated all three vital functions of boiler, cylinder, and condenser into one vessel—very simple, but very wasteful of power. Then we have Thomas Savery and Thomas Newcomen, who reduced waste by separating out the boiler from the cylinder-condenser, but whereas Savery obtained his condenser effect crudely by means of an external tap of cold water, Newcomen had got his effect much less wastefully by using an internal cold spray. And now finally comes James Watt with three separate vessels for each separate function of the boiler, cylinder, and condenser.

There is one very real sense in which it may be said of Watt's design that it was the first steam engine of history. Newcomen's engine was properly called an atmospheric engine, because the actual useful driving-power source was the pressure of the atmosphere, working against an evacuated space on the underside of the piston, produced by the condensation of steam. Watt, on the other hand, was the first to dispense with *air* pressure and to substitute *steam* pressure. This became possible as a direct consequence of the separate condenser, and of the need for efficiency, to keep his cylinder constantly hot. Why not, thought Watt, keep it hot by substituting the steam itself for the atmosphere. Steam is hot and the atmosphere is cold. Let us dispense, then, with the air and surround the cylinder completely with an outer airtight case, filled with steam by direct access to the boiler. The steam would behave just like the air in creating the requisite pressure on the one side of the piston, and would push it to the other end of the cylinder as soon as the space on the other side of the piston was evacuated.

So we get the first real steam engine of history, yet we are still not concerned with steam at high pressure. The theory was complete in terms of steam at normal atmospheric pressure, as long as the "opposition" side was a vacuum. Gradually all these ideas took practical shape, and the result was seen in Watt's first patent in the year 1769. The specification in which he describes his patent marks such an important epoch in the history of steam power that we quote the more important portions of it in full:

> My method of lessening the consumption of the steam and consequently fuel, in fire engines, consists of the following principles:—
> *First*, that the vessel in which the powers of steam are to be employed to work the engine—which is called the cylinder in common fire-engines, and which I call the steam vessel—must, during the whole time the engine is at work, be kept as hot as the steam that enters it; first, by enclosing it in a case of wood or any other materials that transmit heat slowly; secondly, by surrounding it with steam or other heated bodies; and thirdly, by suffering neither water nor any other substance colder than the steam to enter or touch it during that time.
> *Secondly*, in engines that are to be worked, wholly or partially, by condensation of steam, the steam is to be condensed in vessels distinct from the steam vessel or cylinder, though occasionally communicating with them. These vessels I call "condensers"; and while the engines are working these condensers ought at least to be kept as cold as the air in the neighborhood of the engines, by the application of water or other cold bodies.
> *Thirdly*, whatever air or other elastic vapor is not condensed by the cold of the condenser, and may impede the working of the engine, is to be drawn out of the steam vessels or condensers by means of pumps, wrought by the engines themselves or otherwise.
> *Fourthly*, I intend in many cases to employ the expansive force of steam to press on the pistons, or whatever may be used instead of them, in the same manner as the pressure of the atmosphere is now employed in common fire engines [he is here referring of course to the Newcomen engine]. In cases where cold water cannot be had in plenty, the engines may be wrought by this force

of steam only, by discharging the steam into the open air after it has done its office.

Fifthly [here follows a reference to a rotary engine which does not concern us at this stage].

Sixthly, I intend in some cases to apply a degree of cold not capable of reducing the steam to water, but of contracting it considerably, so that the engine shall be worked by the alternate expansion and contraction of the steam.

Lastly, instead of using water to render the pistons and other parts of the engine steam-tight, I employ oils, wax, resinous bodies, fat of animals, etc.

In order to produce an engine model on these lines Watt hired a room in an old pottery. There, through the good offices of Dr. Black, he became acquainted with Dr. Roebuck, who had just set up the famous Carron Iron Works. The expenses to which Watt was put as a result of his experiment considerably impoverished him, and it was not until Dr. Roebuck had agreed to finance him and receive two-thirds of such profits as might accrue that the patent was made possible. Gradually the engine that was to be set up to this specification at an outbuilding adjacent to Dr. Roebuck's house at Kinneil, approached completion: by July of 1769 all was assembled. The boiler was set in position, the great beam was hung, the condenser was rigged with two pumps constructed of tin and hardened lead, and fitted with a strong wooden frame, while the cylinder was of cast iron, with its outer jacket of wood. The workmanship of the cylinder and piston was not considered by Watt to be very good. This was due probably both to defects in the mold and to the boring tool. Consequently, the work of adjustment and fitting took a considerable time. But all was ready by September and the trials began. They were not too satisfactory. The inequalities of fit in the cylinder and piston produced leakages both in the steam and in the oil packing.

He became very despondent, and in his anxiety to relieve Dr. Roebuck of the financial losses that threatened he looked round for other help. This luckily came to hand in the personality of Matthew Boulton, the son of a Birmingham engineer, whom Watt met during a journey to London for the purpose of securing his patent. As a consequence, the famous partnership of Boulton & Watt came into being, to the great enrichment of the profession of engineering. Nevertheless, it was not until 1774 that the first really successful single-acting engine was erected. This is illustrated in the Figure. It will be seen that the top of the cylinder is closed but that the steam from the boiler has access to the upper

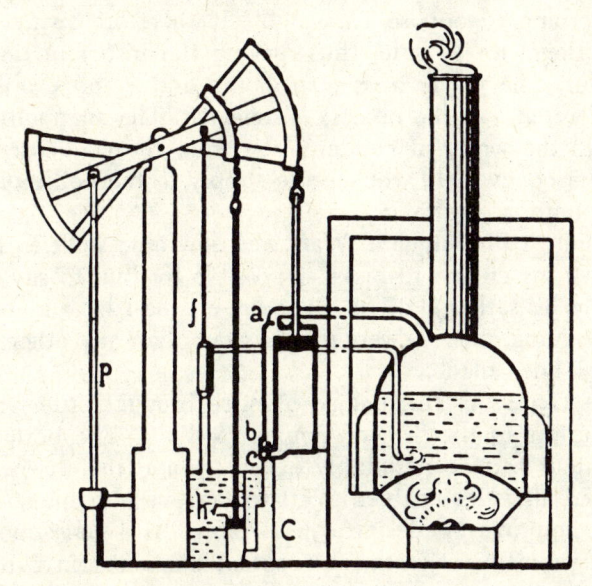

II-1. Watt's first single-acting steam engine.

side of the piston through the admission pipe on the left. This ensured keeping the piston and cylinder hot; nevertheless, as we have already pointed out, the steam on this upper side merely served the same purpose as did the air in the former Newcomen engine. The lower end of the cylinder communicated with the condenser, C. Three valves figure in the design, the steam valve, a, to the boiler, the equilibrium valve, b, between the outer steam jacket and the cylinder, and the exhaust valve, c, to the condenser. The action is, briefly, as follows. At the beginning of the downstroke the exhaust valve, c, is opened to the condenser to produce a vacuum below the piston, and the steam valve, a, is opened to admit steam above it. At the end of the downstroke both these valves are shut, and the equilibrium valve, b, is opened. This gives the steam free access equally in the cylinder to both sides of the piston. As a result the pressure is the same on both sides of the piston. Consequently, the heavy pump rod, in acting as a counterpoise, has nothing to oppose its weight. It therefore swings the beam down on its side, thus driving the piston up in the cylinder. The valve, b, is now closed and a and c opened as before, and so the process is repeated. The air pump discharges the water of condensation from the condenser, C, into the hot well, h, whence the supply to the boiler of the feed pump, f, is drawn.

Living in Birmingham, Watt was now able sufficiently to improve his engine to make it possible for him to say in a letter to his father in 1774: "The fire-engine I have invented is now going, and answers much better than any other that has yet been made."

The success of this engine at once brought orders from all quarters to the firm of Boulton & Watt. The period of the patent had been suitably extended, and the absence of financial anxieties enabled Watt to persevere in his studies and to improve his design considerably. As a consequence, he was able to set forth, in a patent taken out in 1781, a number of important improvements. In particular, these included the double-acting steam engine, by means of which the steam was admitted to each side of the piston alternately, the opposite side being in communication with the condenser. For this purpose he devised a steam chest and slide valve, thus bringing the engine in principle practically to the position it occupies today.

This improvement constituted an advance of considerable ingenuity. It really centered round the problem of the proper utilizing of the expansive force of steam approximately in accordance with Boyle's law. Up till now the piston had been driven for its whole length by the admission of sufficient steam at relatively low pressure. But if this steam is admitted from the boiler at a pressure materially in excess of that of the normal atmospheric pressure, and the supply is cut off when the piston has only gone a fraction of its full distance, the momentum of the piston will carry it on its way, and the steam will expand in accordance with Boyle's law. That is to say, as the stroke continues, the volume of the steam increases and at the same time its pressure will proportionately decrease. Nevertheless, because the steam is initially admitted under considerable pressure, there remains throughout the length of the stroke, after the steam is cut off from the boiler, a sufficient margin of diminishing pressure to keep it greater than that of the atmosphere. In other words, by the device of the "cutoff," the piston is driven along its stroke usefully by the force of the steam, but much less steam is required in so doing.

Originally Watt suggested cutting off the admission of the steam into the cylinder after the piston had completed half its stroke. This meant that the piston would be driven forward the remaining half by the expansive force of the steam. He soon realized, however, that he could work just as efficiently, and much more economically, by cutting off the steam at one-quarter of the stroke.

Watt devised a mechanical attachment to the cylinder, which gave an automatic recording of diminishing pressure on a sheet of paper fitted to a revolving drum, and actuated directly by the steam pressure in the cylinder. It was a simple but ingenious device. In an improved form, this Watt indicator, as it is called, is in general use to this day. It was but a step from this device of the cutoff to the double-acting device of admitting the steam alternately to either side of the piston. Where, previously, there had only been one driving stroke, namely, the downward one (the upstroke being produced by the weight of the pump rod at the far end of the beam), Watt now used a *closed* cylinder, both ends of which communicated with the boiler by the interposition of what is called a steam chest and slide valve.

Watt had in fact been fully alive to this possibility almost from the time he had first thought of his separate condenser. But to work properly it needed the complicated mechanism of the slide valve, which would provide for the automatic timing, first, of the admission of the steam to one end of the cylinder, the other end being closed; and then, alternately, for the "cutting off" of the steam from the open end (at the appropriately predesigned cutoff fraction of the stroke) and the simultaneous opening of the steam outlet at the other end. As we have said, the vessel in which this slide-valve mechanism works is called the steam chest. Watt delayed perfecting it until the time was ripe for this further complication in engine design. The gain in efficiency was tremendous.

Next we come to the introduction of the "rotative" engine. So long as the function of the engine was merely that of pumping water from a lower level to a higher, the simple up-and-down reciprocating motion (which could equally of course be to-and-fro motion, as it is in the horizontal engine) was all that was required. But it was soon realized that, if the steam power of an engine was to be applied to the driving of factory mills, lathes and other machinery plant, then somehow this reciprocating motion of the piston to and fro or up and down would have to be converted into a rotary motion.

Curiously enough, the stock method of today, namely, the use of a crank, was in itself well enough known at the time. The spinning wheel and the foot lathe had long since incorporated this device. But Watt considered this method impracticable for his engine. He was aware of the need for a connecting rod, and for a flywheel to carry the crank and crankshaft over the "dead centers" at the ends of each stroke, but he thought that the strain would be too much for such a mechanism. He was, of course, wrong in this, since in fact the crank does control the stroke of the connecting rod.

One of Watt's ideas for carrying the crank over the dead

centers, namely, the use of a pinion wheel gearing into the crank disk, and having a weight on its edge, was probably "cribbed" by one of his workmen, Richard Cartwright. Watt therefore developed another device known as the "sun and planet" gear. This comprised a "planet" wheel rigidly fixed to the end of the connecting rod, and so called because it was made to rotate by toothed gearing round a "sun," or central wheel, keyed to the shaft that was to be driven. Provided the two wheels are of equal size, the driving shaft will revolve twice for every double stroke of the engine. Nevertheless, all the later forms of Watt's double-acting engine did in fact incorporate the now usual device of the crankshaft and flywheel.

Yet one further adjunct to the Watt engine must be mentioned, the centrifugal "governor." The purpose of this is to ensure the steady motion of the engine, even though the load on it may be altering—so that if the engine is tending to race, the device slows it down, and conversely, if the load is increased suddenly, to tend to jerk the engine down, the momentum of the rotating governor counteracts this and speeds it up momentarily.

This was in effect a double conical pendulum, the rotation of which controlled the speed at which the engine could work. Two heavy iron or brass balls, were suspended from pins from the head of a vertical spindle. The distance of these balls from the spindle was controlled by two jointed rods. These were connected to a collar, whose position on the spindle could be raised or lowered by means of a system of levers connected with the throttle valve of the steam engine. The object of the device was to ensure universal working of the engine. When the engine tended to race, the increased speed would cause the balls to tend to fly outward. This would cause the collar to move upward and so to close the throttle and diminish the admission of steam into the cylinder. As a result, the force on the piston would diminish and the speed of the engine would be reduced. Conversely, if the engine tended to slow down, the balls would sink inward, open the throttle, admit more steam, and so speed up the engine.

Watt first introduced the governor to his engines in 1787. The idea behind it was not new, and he rightly made no claim to be its inventor, nor did he patent it. He did, however, most certainly improve on the existing types and for this is entitled to considerable credit.

By 1787 the Watt double-acting steam engine had taken the form shown in Figure II-2. Steam was passed from the boiler to the cylinder, A, along the steampipe, B, through the throttle valve, C. This valve was in communication with the governor, D. On one side of the cylinder at the upper and lower ends were the hollow steam chests, E. These led to the cylinder by means of a passage in the middle of each. They were each fitted with two valves dividing them into three compartments. The top compartments of both chests communicated with the steampipe, and the lower compartment with the exhaust, or reduction, pipe leading to the condenser. The linkage system was such that the valves worked in pairs, the upper inlet, or induction, valve, F, and the lower outlet, or exhaust, valve, *f*, moving together, and similarly with the upper exhaust valve, G, and the lower induction valve, *g*. The piston, R, fitted accurately into the cylinder by means of a stuffing box. Hence, on opening the valves, F and *f*, steam was admitted above the piston, and at the same time withdrawn from below the piston to the condenser, I. Similarly, by opening the valves, G and *g*, steam was withdrawn from above the piston and admitted below it. All these valves were controlled by a single lever, H, known as the spanner.

The condensing apparatus comprised two cylinders, I and J, both immersed in a cistern of cold water. The pipe, K, was fitted with a rose and conveyed water from the cistern to the cylinder, I, the supply being regulated by a cock. As a consequence, the steam passing into I from the cylinder, A, was condensed. The other cylinder, J, was the air pump, and was fitted with a well-packed piston, L, containing a valve similar to the bucket valve of a common pump. This valve opened upward, drawing off the surplus water as fast as it collected at the bottom of the condenser, I, through the connecting passage and valve at the bot-

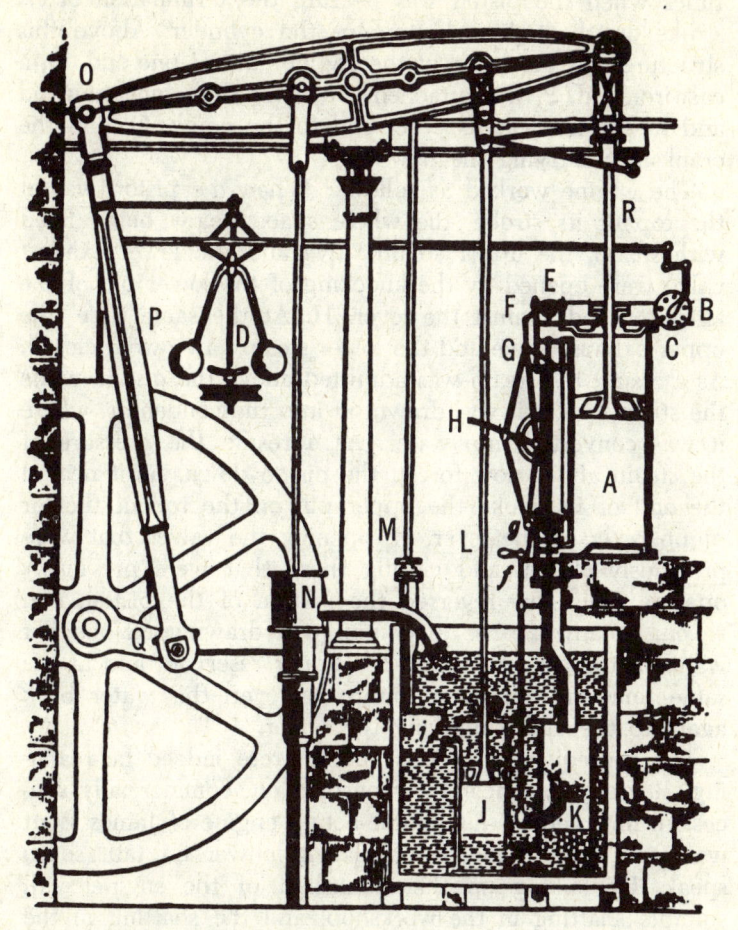

II-2. *Watt's double-acting steam engine.*

tom. The hot-water pump, M, now conveyed this water into the tank that supplied the boiler through the reservoir, *j*. N was a cold-water pump which supplied the cistern containing the air pump, J, and condenser, I, so as to keep it cool. Two pins were placed on the rod of the air pump above and below the lever, H, so as to project outward and strike the lever, H, upward and downward at the proper times when the piston was nearing the termination of its stroke on the top or bottom of the cylinder. Above this structure was the beam of the engine, and at one end, O, a cast-iron rod, P, was attached. This was the connecting rod and it was fixed at its other end to the crank, Q, and the crankshaft carrying the flywheel.

The engine worked as follows: When the piston was at the top of its stroke, the whole space below being filled with steam, the upper steam valve and the lower exhaust valve were opened by the knocking of the lower pin of the air-pump rod against the lever, H. At the same time, the upper exhaust valve and the lower steam valve were closed. As a result, the steam was admitted above the piston, while the steam below it was drawn off into the condenser, where it was converted into water. As a result, the pressure of the steam above now forced the piston down. As it neared the end of its stroke the upper pin on the rod of the air pump reversed the lever, H, opening the valves that were previously closed, and closing those that were previously opened. This now reversed the motion of the piston, and so on. Meanwhile the air pump was drawing off the hot water in the condenser into the upper reservoir, and at the same time the hot-water pump conveyed this water back again to the tank supplying the boiler.

The consequences of this engine were indeed far-reaching. Reasonably efficient in construction, commercially successful in practice, this double-acting engine of James Watt was virtually the means of placing power on tap, so to speak. By linking up the crankshaft of the engine with suitable shafting in the workshop, and the shafting in the workshop to the machinery therein employed, a complete revolution in industry was speedily effected. And because it arrived at a time of sociological importance, when parallel inventions and improvements in the machinery of the textile and other industries were being effected, the whole outlook of industrial life was changed. A new era for industry was at hand.

Despite its superiority over its predecessors, the engine of James Watt utilized steam power in an indirect and highly complicated manner. Its numerous parts and reciprocating action compared unfavorably with such an ancient device as Hero's aeolipile. Numerous inventors of the period, including Watt himself, were fully aware of the fact and applied themselves to finding a remedy along entirely different principles. Watt designed several rotary engines, one of which was "a steam wheel moved by force of steam in a circular channel against a valve on one side, and against a column of mercury, or other fluid metal, on the other side." They were failures. Edmund Cartwright, a Church-of-England clergyman who invented a power loom and a wool-carding machine, and who is generally considered the father of the textile industry, also attempted to design a rotary engine. It remained for Carl de Laval, born in Sweden in 1845, to find the first satisfactory solution. His initial purpose was to invent a power-driven centrifugal cream separator. Successful in this aim, he went on to invent turbines whose efficiency was greatly increased by systems of diverging nozzles which directed steam against the turbine blades. By the end of the century, through the use of constantly higher pressures, the speed of his turbines had increased from hundreds to tens of thousands of revolutions a minute.

Meanwhile the Englishman, Charles A. Parsons, had invented a turbine which was to prove as revolutionary in its effects as the steam engine of Watt. It proved equally important in both the steam and the electrical power industries. Indeed, Parsons' first objective was to build an engine which could drive a dynamo directly—something which required a speed beyond the capacity of the reciprocating engine.

The story of Parsons' contributions to the science of engineering is here told by R. H. Parsons. Like the Huxleys, the Darwins, the Herschels, and the Thomsons, the Parsons were a highly gifted family. Charles was the youngest son of the Irish nobleman, Lord Rosse, builder of the largest telescope of the period. R. H. Parsons, though no relation, was one of Sir Charles' close engineering associates.

SIR CHARLES PARSONS AND THE STEAM TURBINE

R. H. PARSONS

EVEN IF [Sir Charles] Parsons had done nothing more than produce the first practical steam turbine his fame as an engineer would have been secure for all time. By its introduction he exercised an influence upon industry that was comparable only with that of Watt about a century earlier, though vastly more far-reaching in its effects by reason of the wider field open to him.

When Watt built his first condensing steam engines, the operation of pumping machinery for mines was almost the only duty for which engines were required. Towards the end of the eighteenth century steam engines began to take the place of water wheels as prime movers for mills and factories, but as the dynamo had not then been invented, the generation of electricity and all the industrial development that depends on it lay still in the future. Nor was there at the time, except perhaps in the minds of a few enthusiasts, any idea that steam would ever be used for the propulsion of ships. There is no evidence that Watt foresaw the immense field there would be for steam power in marine work, even though his firm of Boulton and Watt constructed the engine that drove Fulton's historic little steam vessel the "Clermont" in 1806. Parsons, on the other hand, commenced his life's work when the two great branches of electrical and marine engineering were already established, each of them offering an unbounded scope for the steam turbine as soon as its practicability could be demonstrated. In another respect also the times were auspicious for him. The reciprocating steam engine, which had held the field unchallenged for a hundred years, had practically reached the limit of its development. The labors of generations of engineers had raised it to a very high degree of excellence, and no further refinement of design was capable of effecting any substantial increase in its efficiency.

Parsons, alone amongst his contemporaries, saw in the turbine principle the means of escape from the limitations of the reciprocating engine, or perhaps it would be more accurate to say that he alone possessed the genius and courage to transform a possibility into a reality. That he foresaw from the outset the wide diversity of the duties to which the turbine could be applied is clear from his earliest patents, which also showed a remarkable understanding of the conditions essential to success in meeting the requirements of each particular case. Problem after problem was solved in the most admirable way, and instead of being merely an ingenious toy, as many people at first considered it, the turbine steadily and surely won recognition as the standard type of prime mover wherever the production of steam power was concerned.

The object that he set himself was that of producing power by utilizing the velocity of a jet of steam, instead of using the pressure of the steam to drive a piston as in the ordinary reciprocating engine. It was evident that a jet of steam could be made to turn a wheel by acting on blades set around its circumference, or alternatively it could be used to develop power by its own reaction when escaping tangentially from an orifice in a rotating wheel or arm. Both devices had already been suggested by innumerable inventors, but the hitherto insuperable difficulty in constructing a practical turbine by either method lay in utilizing the excessive velocity of the steam. Even steam at a comparatively low pressure escaping into the atmosphere may easily be traveling at more than 2500 feet per second, or over 1700 miles an hour, while twice this velocity may be attained by high-pressure steam flowing into a good vacuum. To make use of such velocities effectively in a simple turbine, the blades or other moving elements would have to travel at about half the speed of the steam, for otherwise an undue proportion of the energy of the jet would be uselessly carried away in the steam leaving the wheel. The blade speeds required for efficiency would therefore be so high that they would be prohibited by reason of centrifugal force alone, apart from other considerations.

Parsons could therefore only secure a proper relationship between steam speed and blade speed by reducing the former to a manageable amount. Now the speed of a jet of steam will obviously depend upon the difference of pressure that causes the flow. It occurred to Parsons that he could attain his end by the device of causing the whole expansion of the steam to take place by a series of steps, each partial drop of pressure being only sufficient to generate a velocity that could be efficiently utilized by blades running at a moderate speed. To put this idea into effect he constructed a turbine consisting of a cylindrical rotor enclosed in a casing. The steam flowed along the annulus between the two, parallel to the axis of the machine, and in so doing it had to pass through rings of blades fixed alternately in the casing and rotor. The passages between the blades of each ring formed virtually a set of nozzles in which a partial expansion of the steam could take place. In passing through each ring of fixed blades the steam acquired a certain velocity due to this expansion, and the jets so formed gave up their energy in driving the succeeding row of moving blades. The passages between the latter blades also acted as nozzles, permitting a further partial expansion, so that the moving blades were impelled partly by the "action" of the steam entering them and partly by the "reaction" of the steam leaving them.

In addition to laying down the broad lines necessary to success in the development of the new kind of prime mover, Parsons had many practical problems to solve before his ideas could be embodied in an actual machine. Not only had a suitable form of blading to be invented and appropriate manufacturing methods devised, but the design generally had to conform to conditions quite outside the range of ordinary engineering practice. For example, to obtain the desired blade velocity in the small turbine he first constructed, a rotational speed of no less than 18,000 revolutions per minute had to be adopted. This was over fifty times as fast as the fastest reciprocating engine of the day, and it involved the invention of a new kind of bearing which would permit of a long rotor, inevitably out of mathematically perfect balance, running at such a speed without vibration. Means had also to be provided for the continuous lubrication of these bearings, and a totally new method of controlling the speed of the machine had to be devised. Again, it was realized that the flow of the steam would result in an end thrust on the blading, and to prevent this being transmitted to the bearings, where it might have caused trouble, Parsons neutralized it by the ingenious expedient of admitting the steam midway along the rotor and causing it to flow equally towards each end. His subsequent invention of "dummy pistons" rendered the double flow principle unnecessary for machines of moderate output, but without it the large and efficient high-speed machines of today could hardly be built.

The most obvious field for such a high-speed prime mover as the turbine was in the driving of electric generators, and it was for this purpose that it was originally designed. The dynamos of those days were small machines driven usually at 1000 to 1500 revolutions per minute by a belt from the flywheel of a reciprocating engine. Parsons required a dynamo that could be driven directly by his turbine at a speed of 18,000 revolutions per minute, in order that the combination should constitute a small, simple and self-contained generating unit. None of the established dynamo makers would have considered for a moment the construction of a machine so completely outside the range of previous experience. It must be remembered that at the time electrical engineering was in a very elementary condition and dependent mainly upon empirical knowledge, for two years had still to elapse before Hopkinson propounded the theory of the magnetic circuit and laid down the fundamental principles of the design of electrical machinery. Parsons faced the question with the same boldness that he showed in the design of his turbine and he achieved an equally striking success. Both electrical and mechanical problems had to be solved, for the alternations of magnetism in the core were vastly more rapid than in any machine yet built, while the mechanical stresses to be provided against will be realized from the fact that a centrifugal force of 5½ tons was developed by every pound of metal at the surface of the armature.

It required much persistence on the part of Parsons before his turbo-generators were able to enter their proper field of Central Station work. By 1888, although about two hundred of them were in service, they were employed almost exclusively for ship-lighting duties, and no Electric Light Company had yet taken any notice of them. Parsons therefore decided that he would, himself, have to effect the introduction of the turbine into the industry that it was destined to dominate; so, aided by friends, he founded the Newcastle and District Electric Lighting Co., which began operations in January 1890.

The success of the turbine in saving the chief London Station of the Metropolitan Electric Supply Co. from being shut down altogether in 1894 on account of the nuisance caused by its reciprocating engines attracted general attention and definitely established its footing in the industry. Larger and larger units were continually called for, and with every increase in size the advantages of the turbine became more apparent. Parsons would have attained high fame for his electrical work alone had not this been overshadowed in the minds of the public by the spectacular developments of his steam turbine. By 1900 he was building generating sets of 1000 kW. capacity, while in 1912 he undertook to build a turbo-alternator with an output of 25,000 kW., by far the largest and most efficient generating unit in the world at the time. This machine was installed in the Fisk Street Power Station of the City of Chicago, and it proved so successful that in 1923 Parsons was entrusted with the contract for a unit of 50,000 kW. for the same city. He lived to see an output of more than 200,000 kW. delivered by a single turbo-generator, and the reciprocating steam engine completely superseded by the turbine for Central Station work.

The growth of electricity supply consequent upon the invention of the turbine created a demand, not only for larger generating units, but also for higher transmission voltages, in order that more extensive areas might be economically served. In the early days the practice had been to generate at about 2000 volts, and to step up this pressure when required by means of transformers. Ferranti had given a lead in the direction of higher generating voltages in 1889 by designing large slow-speed alternators to generate single-phase current at 10,000 volts for his famous Deptford Station. These machines were, however, recognized as exceptional and they had little or no influence on the industry generally. The first real advance towards modern conditions was made by Parsons in 1905 when he supplied a pair of 1500 kW. turbo-alternators generating at 11,000 volts to the Frindsbury Power Station in Kent. Once it had been demonstrated that high-speed alternators could be safely constructed for this voltage, it soon became a usual generating pressure and remained so for many years. As before, when higher pressures were required for transmission they were obtained by the use of transformers, which were commonly attached permanently to the machines they served. There was, however, to Parsons' mind, something illogical in generating at 11,000 volts or thereabouts when the whole of the current might have to leave the Station at a higher voltage. He therefore attacked the problem with his usual energy and insight, with the result that in 1928 he produced a 25,000 kW. turbo-alternator designed to generate directly at 36,000 volts. This was installed in the Brimsdown Power Station the same year. The windings of the machine were constructed in accordance with an entirely new principle, which made it possible to generate at 36,000 volts without submitting the insulation of the windings to any greater electrical stress than is usual in an ordinary 11,000 volt generator.

Enough has been said to indicate, in some small measure, how much the electrical industry owes to Parsons. He not only provided it with the turbo-generator, but led the way for more than a generation in every important development of Power Station machinery. Typical of his engineering courage was the jump from 350 to 1000 kW. in 1900, and the still more spectacular leap from 6000 to 25,000 kW. in 1912. He was an equally great pioneer in all matters pertaining to efficiency. As long ago as 1900 he made the first practical experiments with regard to the reheating of steam in the course of its expansion.

Although the technical merits of Parsons' work in the development of turbines and electrical generators can only be fully appreciated by experts, the benefits that have accrued from it are obvious to all. It is sufficient to contemplate the part played by electricity in our domestic and industrial well-being to realize how greatly we are dependent upon a cheap and abundant supply. This has come to be regarded as one of the necessities of civilized life, and it is very certain that the service we now enjoy would have been utterly impossible without the turbo-generator.

Although the steam turbine finds its greatest field of usefulness on land in Power Stations and in driving the electrical generators in private plants, it has many other industrial applications. Turbines were employed at a very early date for driving centrifugal pumps, fans and blowers, all of which were naturally suitable for direct operations by a high-speed prime mover. Parsons also realized that if a turbine were driven by power, instead of being used to produce it, the machine could be used as a compressor. Many compressors for air and gas were constructed by him on the principle of the reversed axial-flow turbine.

The perfection of toothed gearing, to which Parsons contributed so greatly by his invention of the "creep" system of cutting the teeth of gear-wheels, opened up the whole industrial field to the turbine, as it was then no longer confined to the driving of such high-speed machinery as could be directly coupled to it. It was successfully applied even to the driving of steel rolling mills and other duties of a similarly exacting nature, and was often used to replace ordinary steam engines for driving the main shafts of factories. Even when reciprocating engines were retained, the ability of the turbine to work with steam at very low pressures was frequently taken advantage of by installing turbines to develop extra power from the exhaust steam of the engines which had hitherto been blown to waste. In all these developments Parsons took a leading part, providing turbine machinery to meet the most diverse requirements and thereby enabling factories and industrial undertakings to produce their own power much more cheaply and efficiently than before.

The use of the steam turbine for the propulsion of ships was amongst the claims made by Sir Charles Parsons in his original patent of 1884, but he confined his energies at first to the task of establishing the position of the turbine on land, and it was not until 1894 that he took steps to apply it to marine duties. He decided to establish a separate organization to deal with the special problems involved in marine propulsion. This company, which became known later as The Parsons Marine Steam Turbine Co. Ltd., proceeded immediately with the construction of a little vessel whose fame is now historic. This was the "Turbinia" with a length of 100 feet and a displacement of 44 tons. After much experimental work with her propellers, the "Turbinia" attained a speed of 34 knots, which was a very remarkable achievement, since the fastest destroyers of the time could hardly exceed 27 knots. The fact that the steam turbine was inaugurating a new era in marine practice was brought home to the public in an unmistakable manner at the great Naval Review held in 1897 to celebrate the Diamond Jubilee of Queen Victoria. A vast fleet, representing not only the might of the British Navy but the sea-power of other leading nations as well, was assembled off Spithead when the little "Turbinia," with Parsons himself in control of the machinery, created a sensation by racing down the lines of warships at a speed obviously greater than that of any other vessel afloat. The Admiralty could not ignore such a demonstration, and entrusted Parsons with the construction of a 30-knot turbine-driven destroyer, H.M.S. "Viper," but so grudgingly was the order given that Parsons and his associates were required to deposit a sum of no less than £100,000 as a security, in case the vessel should not come up to expectations. These, however, were more than fulfilled, the "Viper" attaining a speed of over 37 knots when officially tested over the measured mile with turbines developing 12,000 h.p.

In 1901 the first turbine-driven passenger vessel, the "King Edward," was built for service on the river Clyde. This was followed by the "Queen Alexandra" for the same duties, and within the next year or two, turbine propulsion had also been adopted for the cross-channel boats "Queen" and "Brighton." Meanwhile, in order that the advantages of turbines for warships should once more be demonstrated to the Naval Authorities, The Parsons Marine Steam Turbine Co. laid down another turbine-driven destroyer which was eventually acquired for the Fleet under the name of H.M.S. "Velox." The Admiralty now began to take the turbine more seriously, and when, in 1902, orders were placed for four 3000-ton cruisers, it was decided that one of them, H.M.S. "Amethyst," should be fitted with turbines in order that a comparison might be made between her performance and that of the three sister vessels equipped with the usual reciprocating engines. The results were so conclusively in favour of the "Amethyst" that the last prejudices against turbine machinery in the Navy were overcome and the way was open for its general adoption.

The first turbine vessel to cross the Atlantic was the steam yacht "Emerald," built in 1903 to the order of Sir Christopher Furness. The Cunard Company followed with a 30,000-ton liner, the "Carmania," which once more demonstrated the superiority of the turbine by proving, on her trials in 1905, fully a knot faster than her sister ship the "Caronia," equipped with reciprocating engines. By this time Parsons had won his battle for the recognition of the turbine at sea.

In 1904 the British Government came to an arrangement with the Cunard Company, under which the latter should construct two new liners with an average speed of at least 24·5 knots. The two new liners, the "Lusitania" and "Mauretania," were launched in 1906 and went into service the following year. The vessels were practically identical in design. The "Mauretania" had a displacement of 38,000 tons, and obtained a speed of 26·04 knots on her 48 hours' full-power trials with her turbines developing 70,000 horse-power. She captured the "Blue Ribbon of the Atlantic" for the fastest crossing and held this honor for nearly a quarter of a century. Her sister vessel, the "Lusitania," will be remembered as having been sunk without warning by a German submarine in 1917 with the loss of over a thousand passengers and crew. In the Royal Navy H.M.S. "Hood" was constructed during the last war with turbines of 150,000 h.p. and even this power has been considerably exceeded in recent Atlantic liners such as the "Queen Mary" and "Queen Elizabeth." The subsequent introduction by Parsons of gearing between the turbines and the propellers was another great step in advance, for it not only diminished the size of the machinery and increased its efficiency, but it enabled the ordinary cargo vessel to profit equally by the employment of turbines.

The turbine, however, has never altogether succeeded in ousting the well-tried marine engine from slow-speed cargo vessels. On the contrary it did much to give the engine a new lease of life, for by the addition of a turbine to develop power from the exhaust steam of the engines, the efficiency of the machinery was considerably increased.

The insistent desire of shipowners for an even greater economy of fuel led to the development of the marine oil engine, which, after the end of the War of 1914–1918, began to challenge the supremacy of turbine machinery, particularly for mercantile vessels of slow and moderate speed. The turbine, however, had the advantage of much greater mechanical simplicity, and Parsons sought to bring its fuel consumption more nearly into line with that of the oil engine by urging the adoption of higher steam pressures and temperatures at sea. Although marine engineers had been converted to a belief in turbines and gearing, they had always been conservative in the matter of boiler practice. In 1926 steam conditions in the mercantile marine did not much exceed a pressure of 200 pounds per square inch and a temperature of 500 degrees Fahrenheit. In Naval work matters were somewhat better, but not much, as the steam pressure in warships was only about 275 pounds. Practice on land was very much more advanced. At that time many Central Stations were already using steam at 500 or 600 pounds pressure, superheated to 750 degrees Fahrenheit or over, while in some cases pressures of the order of 1400 pounds per square inch had been adopted. Parsons felt very strongly that marine engineers ought to take advantage of the economies resulting from higher pressures and temperatures. Knowing that a practical demonstration was the surest and quickest way of convincing the sceptics, he arranged for the equipment of a small passenger vessel, the "King George V," with geared turbines of 3500 h.p. to work with steam at 550 pounds per square inch, superheated to 750 degrees Fahrenheit. The steam was supplied by water-tube boilers of the Yarrow type. The "King George V" was the pioneer of high-pressure steam at sea. Although the installation was a comparatively small one, and the conditions of a river steamer making short trips with frequent stops were not the most favorable for the experiment, the machinery fulfilled the expectations of its designers, the full-load trials made after a short period of commercial service showing a steam consumption of only 8·01 pounds per shaft horse-power hour of the turbines. This enterprise of Parsons once more opened up a new field for marine engineers, and higher pressures at sea soon became general.

The application of the steam turbine to marine propulsion gave rise to many incidental problems, by the solution of which Parsons made notable contributions to the progress of marine engineering. One of the earliest difficulties he encountered was due to the high speed of the propellers. The first machinery of the "Turbinia" consisted of one turbine driving a single propeller at 2000 r.p.m. The results of the trials were disappointing. Different designs of propeller were tried but the best speed that could be obtained was only about 20 knots. It was clear, either that the turbine was not developing its rated power, or that the efficiency of the propeller was extremely low. To settle this question Parsons devised a special apparatus to measure the torque exerted by the turbine on the propeller shaft. This instrument was the prototype of the modern torsion meter, and by its use he assured himself that the fault was in the propeller and not in the turbine. About the same time similar difficulties in obtaining the anticipated speed were experienced in a new class of very fast torpedo boats which were fitted with reciprocating engines. Both Parsons and the Naval Authorities arrived at the same conclusion, namely, that the trouble was caused by the inability of the water to follow the rapidly moving propeller blades, so that a vacuous space was left behind the blade tips, with a consequent loss of propulsive power. This phenomenon, now known as "Cavitation," is also liable to occur in centrifugal pumps and water turbines when conditions are favorable to it. Parsons met his immediate difficulties by providing the "Turbinia" with three shafts each carrying three propellers so that the whole propulsive power was divided among nine propellers. With this alteration the vessel attained a speed of over 34 knots.

Many men would have rested content to have successfully circumvented their difficulty, but Parsons realized the importance of a thorough investigation of the whole question of cavitation, as this was clearly going to be a matter of concern to designers of high-speed vessels. He therefore constructed a tank with glass sides in which a model propeller could be run at high speeds. The propeller was strongly illuminated by intermittent light, the speed of the flashes being regulated in accordance with the revolutions of the propeller so that the blades could be made to appear stationary or only revolving very slowly. It was recognized that cavitation would be favored by working with water near its boiling point, so the first experiments were made with hot water. It was found, however, more convenient to attain the same result by maintaining a vacuum above the water in the tank, and the nature of cavitation was exhaustively studied in this manner. The knowledge gained by these investigations led to great improvements in the design of high-speed propellers, and the methods of study initiated by Parsons have since become generally adopted.

Closely allied with the phenomenon of cavitation is that of the erosion of propeller blades, although the connection between the two was not at first realized. Erosion had become such a serious problem that in 1915 the Admiralty appointed a Committee to report on the subject. In view of Parsons' experience of propeller design, he was requested to serve on the Committee, and it was he who suggested that the erosion was probably a secondary effect of cavitation. His view, which is now generally accepted, was that the vacuous spaces typical of cavitation were continually collapsing, causing a hammering by the water on the metal of the propeller. This hammering might easily attain a destructive intensity owing to the absence of any appreciable quantity of air or gas in the cavity to soften the blows.

It was typical of Parsons that he would accept no theory, not even his own, that could not be supported by experiment, so he set himself to test his idea. The method he adopted was as simple and direct as it was ingenious. He made a hollow brass cone with a small hole in its apex which could be closed by a plate of the metal to be operated on. This cone was held face downward in a tank, and when filled with water it was forced suddenly downward until arrested by a rubber cushion on the bottom of the tank. The resilience of the rubber permitted the water in the cone to continue its downward motion for a moment after the cone had stopped, thus causing a vacuous space to occur at the top of the cone. This space immediately collapsed, and the returning water was found to strike the plate with a force often sufficient to puncture it. Pressures as great as 140 tons per square inch were obtained in this way, and the results of the experiments left no doubt that the damage met with in propeller blades could be fully accounted for by the hammering action consequent upon cavitation, as suggested by Parsons.

The steam turbine is essentially a high-speed prime mover, and it therefore shows to its best advantage when directly coupled to machinery that can be run at the economical speed of the turbine. This speed is, fortunately, suitable for a large range of electrical machines, but the smaller sizes of alternators and most continuous current generators require to run at less than the optimum turbine speed, which indeed may be altogether too high to make direct driving advisable or even practicable in many instances.

The obvious way of arranging for the speed of the turbine to be independent of that of the driven machinery is by interposing speed-reducing gear between the two. The immense fleets of slow-speed tramp steamers and cargo ships on all the seven seas were unable to benefit from turbine propulsion because the normal speeds of their propellers were too far below the rotational speeds desirable for turbines.

The reason for the discrepancy between the most efficient speeds of a turbine and a propeller lies in the enormous difference in the density of the media—steam and water—in which they are respectively working. A turbine can only be made to run slowly with efficiency, either by constructing it with a very large diameter in order to maintain the peripheral speed of the blades, or by using a very large number of blade rows so as to reduce the steam velocity per stage. In either case the dimensions become excessive. Parsons realized that the only real solution to the problem was to be found in providing some connection between the turbine and the propeller shaft which would enable each to run at the speed most conducive to efficiency. He commenced by carrying out exhaustive experiments to determine what tooth-speeds could be employed and what power could be transmitted consistently with safety and durability. The results were so encouraging that new possibilities were opened up for turbine machinery both on land and sea. To make a practical test of the use of gearing in marine work, The Parsons Marine Steam Turbine Co. purchased in 1909 an old cargo vessel, the "Vespasian," of 4350 tons displacement, and replaced her 750 h.p. triple expansion engines by geared turbines. The success of the experiment has become historical. With the same boilers and steam pressure, the substitution of geared turbines for the original machinery resulted in a reduction of the fuel consumption by 15 percent. The gearing worked perfectly, and after the "Vespasian" had completed several years of commercial service, her hull, which was then worn out, was broken up and the turbines and gearing transferred to another vessel.

By 1919, or only ten years after the first experiments with gearing in the "Vespasian," it was estimated that no less than 18,000,000 h.p. were being transmitted through gearing in warships and merchant vessels, and as much as 25,000 h.p. had been transmitted by a single gear-wheel.

Simultaneously with the application of gearing to marine purposes, an equally bold departure was made by Parsons in land practice. He supplied a 750 b.h.p. turbine running at 2000 r.p.m. for the onerous duty of driving a rolling mill for the production of ships' plates. The rolls had to run at 70 r.p.m., and this speed was obtained by the interposition of double reduction gearing between the turbine and the mill. The plant proved a most gratifying success, and there could no longer be any doubt that the use of mechanical gearing would enable the turbine to drive the reciprocating engine from almost its last strongholds.

In order that gear-wheels should work quietly and without deterioration under the conditions of speed and power imposed by their new duties, it was of course essential that their teeth should be extremely accurate both as to form and pitch. In the ordinary method of gear cutting, every error of pitch that may exist in the master wheel of the gear-cutting machine will necessarily be reproduced in the wheel being cut. No master wheel can be mathematically perfect, and the accuracy desired by Parsons was greater than any ordinary gear-cutting machine could provide. He therefore turned his attention to the production of better gears, and to this end he devised what is known as the "Parsons Creep Mechanism." By this mechanism the work table of the machine was caused to rotate slightly faster than the master wheel, with the result that any errors existing in the latter were distributed spirally round the wheel being cut, instead of being concentrated at one part of the circumference. The consequence was that the unavoidable defects of the master wheel were, for all practical purposes, completely eliminated in the work.

This method of "creep-cutting," invented by Parsons in 1912, created an entirely new standard of accuracy for mechanical gearing, and made it possible to produce gear-wheels that could be relied on to transmit any desired power with quietness and durability. Thenceforward the turbine was free from all limitations imposed by the speed of the driven machinery, for each could be run at its most efficient rate, the connection between the two being made by appropriate gearing. Geared turbines were soon employed as the propelling machinery for steam ships ranging from slow cargo vessels to the fastest warships and liners, for even in the case of fast ships it was recognized that high-speed turbines and gearing were more economical in service than direct-coupled machines running at a speed dictated by the requirements of the propeller. The improvement brought about by gearing may be illustrated by the following comparisons. With the early direct-coupled marine turbines the steam consumption was about 15 to 16 pounds per shaft horse-power hour, while by 1923 geared installations could operate with a consumption of less than ten pounds of steam for the same power, which could be reduced to eight pounds or less if the steam was superheated. If we take into account the simultaneous increase of the efficiency of the propeller due to its lower speed, it is fair to say that by the introduction of gearing, Parsons effected a further economy in fuel comparable with that originally brought about by the application of the turbine to marine work.

Whatever aspect of Parsons' work is considered, one is struck by the extraordinary ability, energy, and courage which he brought to bear on every problem in hand. The results he achieved have revolutionized whole branches of engineering and exercised a profound influence upon the structure of civilized life. Yet the means by which his mind was guided are, for the most part, as inexplicable as manifestations of genius always must be. His success in developing the steam turbine was certainly not the outcome of any theory of thermodynamical processes, for such theories were at the time in a rudimentary state. Nor was he ever known to make the slightest use of the formal procedure of mathematical reasoning, though his mathematical attainments were high enough to have gained special distinction for him at Cambridge. His intuition served him as an infallible guide in design, and elementary arithmetic sufficed for such calculations as he ever made. He was a profound believer in experiment, and had the faculty of devising simple experiments that would give him exactly the data he sought. As an example of his skill, he determined the power that would be required to drive the "Turbinia" at full speed by towing a small model of the hull across a pond by means of a fishing rod, and his forecast proved to be practically exact. His dexterity as a mechanic was extraordinary. He delighted in the making of models, not for show, but for their utility in the demonstration of some principle in which he was interested at the time. Any material that might be at hand was made use of, and the result was always something that would work.

On Parsons' engineering courage there is no need to enlarge. A man who would take the responsibility for the 70,000 h.p. turbines of the "Lusitania" and "Mauretania" when the most powerful turbines afloat did not exceed 14,000 h.p. and who would construct a 25,000 kW. turbo-alternator for a foreign Power Station when his previous largest unit had a capacity of no more than 6000 kW. was certainly not lacking in boldness. There was, however, nothing of recklessness in his nature. His courage was tempered with prudence and he would never allow himself to be persuaded into any undertaking that he felt instinctively to be unsound. But once he had satisfied himself that any desirable object could be achieved, the magnitude of the step required for its attainment never deterred him for a moment. He was always anxious to make advances in practice, and his enterprise was boundless. His whole career was indeed a continuous progress into uncharted fields, undaunted by fears of failure and guided to success by an incomparable genius.

One of the most delightful features of Parsons' character was his extraordinary modesty. Even when at the zenith of his fame he seemed unable to realize that his achievements had been exceptional or that his ability was anything out of the ordinary. This natural humility of his disposition led him to expect other people to possess an insight equal to his own, and he was always ready to listen to any reasonable argument and to discuss it on terms of equality, though he was capable of an occasional abrupt explosion of impatience with stupidity of thought or action. His manner was kindly, courteous, and considerate to all, and his public and private benefactions were many. He contributed freely of his knowledge to the proceedings of scientific and technical societies, and many of them, in return, conferred upon him the highest distinctions in their power.

Sir Charles Parsons died on February 11, 1931, while on a cruise in the West Indies, and there then passed away one of the greatest engineers of all time. He was in his 77th year, having lived to see the fruit of his labors in the complete transformation of the methods of producing power from steam, both on land and sea. To few men is it given to accomplish so much for the material benefit of humanity, or to set so high an example to those that come after.

While Savery, Newcomen, and Watt were investigating the power of steam, another and parallel series of experiments, seemingly trivial in nature but destined to have even greater influence on the development of civilization, were taking place. It had been known since ancient times that certain substances such as amber would attract light objects like paper or feathers if rubbed. This interesting but seemingly unimportant phenomenon attracted the attention of a group of natural philosophers and a small store of knowledge about it was obtained. Von Guericke invented a machine for generating a continuous supply of electricity; Du Fay discovered that static electricity could be either positive or negative; Musschenbroek invented the Leyden jar, in which electricity could be stored; Franklin proved that lightning was electrical in nature.

In 1786, Luigi Galvani of Bologna, through one of those accidental observations which have played so large a part in the advancement of science, established the existence of "galvanic" electricity. Dissecting a frog's leg with a steel knife which came into contact with a friction machine, he noticed that the muscle contracted. Later he obtained the same reaction by touching the leg with a rod composed half of zinc and half of copper. He came to the conclusion that the force resided not in the metals but in the leg and that in electricity he had discovered the vital force which natural philosophers had been pursuing for centuries. His hypothesis did not jibe with the fact that the force was not apparent when a rod consisting of a single metal was used.

It was Alessandro Volta of Pavia who came to the conclusion

that it was the combination of metals which caused the phenomenon and who by inventing the Voltaic pile—succeeding layers of zinc, wet paper, and copper, with wires attached to the lowest zinc and highest copper layers—produced the first current, as opposed to static, electricity in history.

Still another clue to the nature of this mysterious substance was provided by Hans Christian Oersted and André-Marie Ampère, who almost simultaneously demonstrated that magnetism and electricity were related. Following this observation, as Dr. Sharlin points out, the next step was the inspired research of Michael Faraday, whose newborn baby, the generation of electricity through magnetism and motion, now runs the dynamos of the world. Like Watt, Faraday was a poor and uneducated youth who won the patronage of a professional scientist—in this case Sir Humphrey Davy of the Royal Institution. With the researches of these two men, a whole period in world history, which may be said to have existed from the earliest times, came to an end. The age of modern power began.

In this article, Dr. Sharlin explains not only Faraday's crucial experiments but the important subsequent developments which contributed to the building of the electrical power industry.

FROM FARADAY TO THE DYNAMO

HAROLD I. SHARLIN

IN TEN INSPIRED DAYS during the fall of 1831 Michael Faraday discovered electromagnetic induction, found essentially all the laws that govern it, and built a working model of an electric dynamo. Then he moved on to other research. "I have rather," he wrote, "been desirous of discovering new facts and new relations dependent on magneto-electric induction, than of exalting the force of those already obtained; being assured that the latter would find full development hereafter."

Full development took a long time. Not until the 1880's were Faraday's theories, and the technical clues he provided, embodied in really efficient electric generators. The 50 years in between constitute an engineering gap: the period that separates a piece of basic research from its practical application. The history of the ten days and the half-century furnishes an excellent example of the process by which science passes into technology.

The process is typically divided into two phases. During the first, a fundamental discovery has been made but no one sees a possibility of using it. The new field attracts only the pure scientist and the dabbler in curiosities, neither one aiming at a practical goal. This is the time when theory is ahead of application.

Then there appears a technological niche for the discovery, usually as a result of advances in some collateral area. The niche may open up in months or in years—perhaps never. When it does, the engineer, the inventor, and the businessman enter the arena. But now they are likely to find that theory is inadequate to their purposes or that they do not understand it well enough to use it. In this second phase application suffers from inadequate theoretical support. Catalyzed by economic incentive, the pace quickens. A growing technology begins to contribute to science, as well as the other way around, and theory is extended and deepened; devices based on what man thinks he knows about nature are mirrors of truth that bring him closer to understanding the material world. Eventually the interaction closes the engineering gap.

When Faraday began, he knew just what he was looking for. He undertook his experiments with the explicit "hope of obtaining electricity from ordinary magnetism." That hope was prompted by Hans Christian Oersted's demonstration that magnetism could be obtained from electricity. Oersted had been trying to find out if electric current, made available by the recent invention of the chemical battery, exhibited the same attractive power as static electric charge. In 1820 he found that while a current flowing through a wire does not

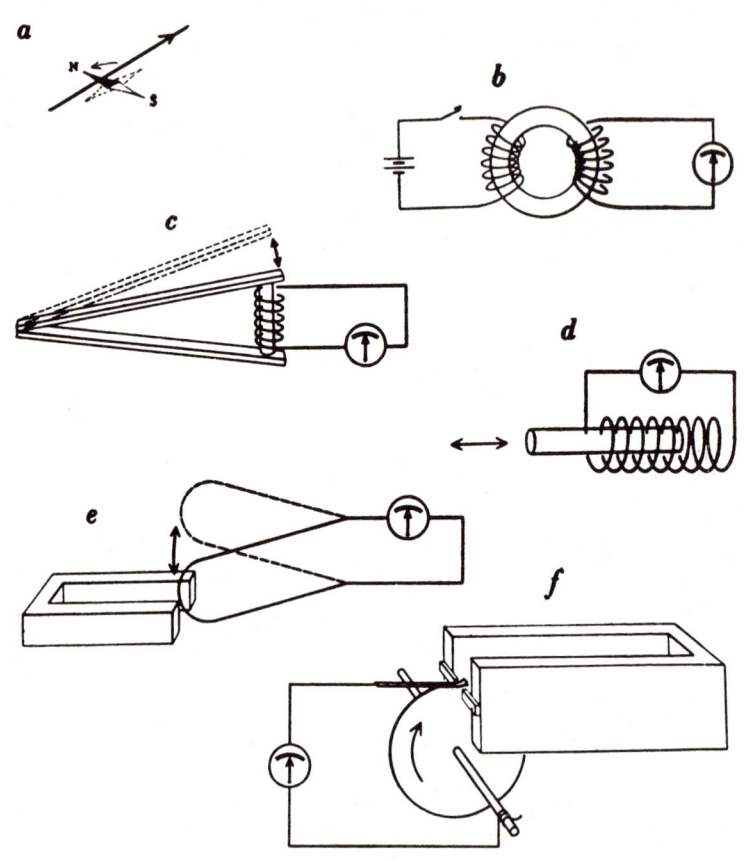

II-4. FARADAY'S EXPERIMENTS were prompted by Oersted's discovery that a current-carrying wire made a compass needle near it (a) swerve at right angles to the wire, showing that electricity produced magnetism. Faraday sought to show that magnetism could produce electricity. He found first that when a coil on one side of an iron ring (b) was connected to or disconnected from a battery, a surge of current was sent through a coil on the opposite side of the ring. Then he found that the same effect could be obtained by making or breaking the magnetic contact between two bar magnets and a coil wound on an iron core (c), by thrusting a magnet into a coil of wire or withdrawing it (d), or simply by moving a loop of wire up and down in a magnetic field (e). Finally he rotated a copper disk between the poles of a powerful magnet and found that a steady electric current was induced across the disk (f).

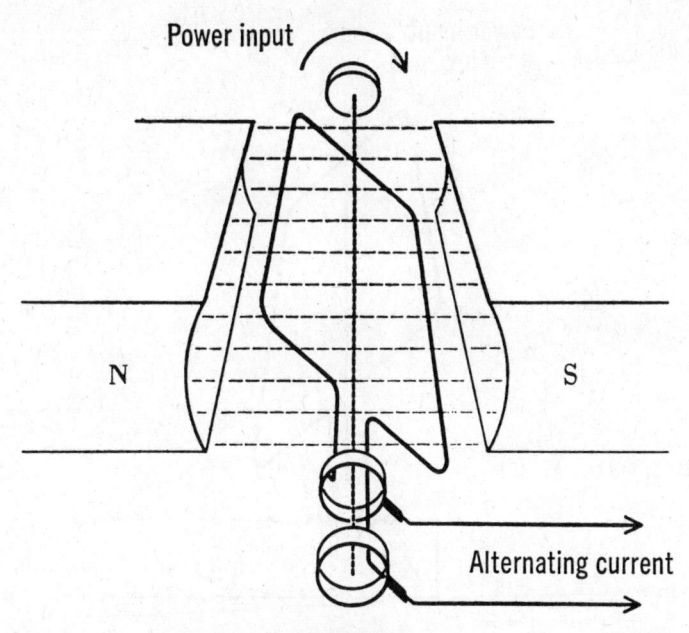

II-5. DYNAMO PRINCIPLE *is illustrated. A loop of wire (the armature) is rotated so as to cut the lines of force between magnetic poles. A current—clockwise in this case—is induced in the loop, which is connected to brass slip rings, and the current is led to the external circuit by two brushes. The current alternates because it reverses directions as the two sides of the loop cut the magnetic field first in one direction and then in the other.*

attract objects, it does cause a magnetic needle to line up perpendicularly to the wire.

As early as 1822 Faraday wrote in his notebook: "Convert magnetism into electricity." The same idea occurred to the two great French physicists André Marie Ampère and Dominique François Arago, but both soon decided there was nothing in it. There was no way to arrive at electromagnetic induction by reasoning from the scanty theory of the time. The effect would have to be discovered by experiment, and Faraday was the supreme experimenter.

Four times between 1822 and 1831 he tried and failed.

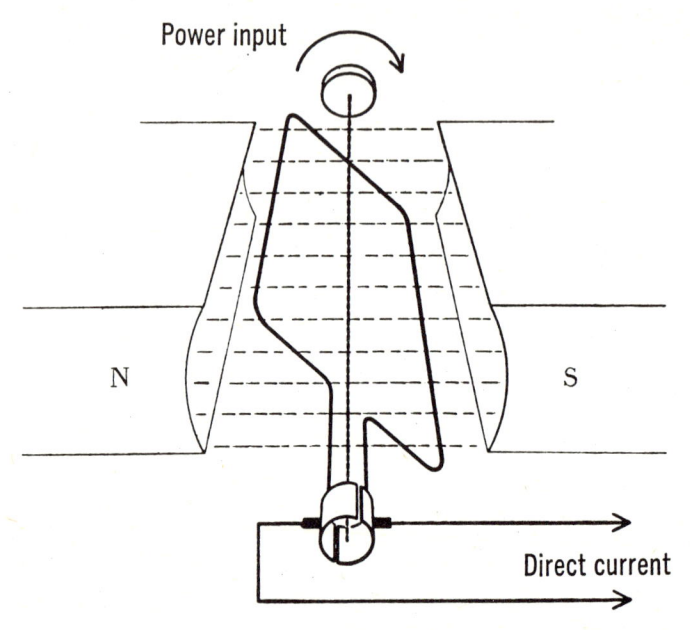

II-6. *DIRECT-CURRENT DYNAMO is made by substituting a commutator for slip rings. The commutator, a ring divided into two segments, switches the sides of the loop from brush to brush so that the current flowing through each brush always goes in the same direction.*

On August 29, 1831, he began his fifth attempt and was rewarded almost at once by the happy accident that every experimenter hopes for. He had wound two coils of wire on opposite sides of an iron ring, insulated from each other and from the ring. A battery sent current through the first coil, magnetizing it, and a galvanometer was connected to the end of the second. As in all the previous trials, no current was detected in the second coil. But then Faraday noticed that whenever the battery was connected to or disconnected from the first coil, the galvanometer indicated a momentary current. He had at last found the key: a *change* in the magnetic field created by the first coil produced a current in the second.

Faraday immediately set out to investigate all possible types of this "transient effect," as he called it. He wound a coil of wire around a straight core and placed the core between two bar magnets arranged to form a V. When he pulled the magnets away, a current flowed through the coil. "Hence," he noted in his diary, "distinct conversion of Magnetism into Electricity." Similarly, he induced a current by thrusting a bar magnet into a coil of wire and obtained a current in the opposite direction by withdrawing the magnet. And he reduced his apparatus to its fundamentals when he induced a current in a simple loop of wire merely by passing it through a magnetic field.

All these experiments produced intermittent surges of current, lasting only as long as the relative motion of conductor and magnetic field. Faraday now arranged for continuous motion by rotating a copper disk between the poles of a permanent magnet. A wire around the axle of the disk ran to a galvanometer, and another wire led back from the meter to a metallic conductor held against the rim of the disk. As long as the disk was turned, the galvanometer indicated a continuous current. "Here therefore," Faraday wrote, "was demonstrated the production of a permanent current of electricity by ordinary magnets." He called the device a "new electrical machine" and suggested that its power could be increased by using several disks. Then he dropped the matter.

Faraday's experiments and his observations on them actually contained a number of clues to effective generator design. Had they been recognized, much of the trial, and more of the error, of the next half-century would have been avoided.

The most important clues were contained in his general statement: "If a terminated wire [*i.e.*, one forming part of a complete circuit] move so as to cut a magnetic curve, a power is called into action which tends to urge an electric current through it." This revolutionary idea of magnetic curves or lines of force was not accepted by most of the physicists of the time. It was not until James Clerk Maxwell published his mathematical interpretation of Faraday's model in 1864 that the idea took hold.

But Faraday had already shown in 1831 that, in each of his methods of producing electricity from magnetism, the cutting of lines of magnetic flux by a conductor is the crucial factor. This was true whether he changed the field (by connecting or disconnecting the battery), moved the magnet or moved the conductor. He had discovered the principle that came to be called Faraday's law, which states that the voltage induced in a conductor is directly proportional to the rate at which the conductor cuts lines of magnetic flux. To maximize the rate, the conductors in an ideal generator should pass through the field at right angles to its lines of force. This is perfectly obvious, but only to someone who visualizes the magnetic field as being made up of lines of force. Those who followed Faraday did not, and as a result an efficient armature did not appear for many years.

Another clue lay in the fact, duly recorded by Faraday, that coils wound on an iron ring gave an induced current "far beyond" that obtained from coils on a wooden core. The current was stronger because the iron provided a better magnetic circuit, concentrating the flux so that more lines of force passed through and cut the second coil. Neither Faraday nor his successors realized this, and in the early development of the dynamo the question of the magnetic circuit was ignored. It was simply adapted to fit each change of shape in the armature, sometimes by chance increasing the flux cut by the conductor, but just as often decreasing it.

Having used both permanent magnets and electromagnets in his experiments, Faraday remarked on the "similarity of action, almost amounting to identity, between common magnets and either electro-magnets or volta-electric currents." Yet he continued to distinguish between the two sources of magnetism. And the early builders of generators for some reason used clumsy permanent magnets exclusively, although electro-magnets are lighter and more powerful. It was not until the 1860's that electro-magnets were generally adopted.

With the conclusion of Faraday's 1831 experiments the

first phase in the development of dynamo technology opened. The basic discovery had been made; there was theory, but no immediate interest in applying it. Electromagnetic induction seemed a far less powerful source of current than the chemical battery. More important, there was no apparent use for large currents of electricity and no incentive to develop machines to generate them.

As always, there were tinkerers. In 1832 Hippolyte Pixii exhibited a machine based on Faraday's principles in Paris. Producing very little power, it was in effect no more than a model of a generator. The device had stationary coils and a hand-driven rotating horseshoe magnet; its output was limited by the weakness of the magnet and the energy available in the operator's arm. Even so, Pixii could have increased its power substantially if he had understood the importance of the relation between magnet and conductors. He wound his conductor coils on two bobbins, a neat way of getting a long length of wire into a small space. At best, however, only a small proportion of the wire can ever be perpendicular to the field in this arrangement.

Pixii's first model produced alternating current as the rotating field cut the conductors first in one direction and then in the other. Alternating current seemed altogether pointless, and at Ampère's suggestion Pixii equipped his second version with a commutator so that it would deliver direct current, as a battery does. (A commutator is a rotating switch that reverses the connection between the armature winding and the outside circuit each time the current changes direction in the winding, so that the external current always flows in the same direction.) Actually, alternating current is better suited to electric power transmission, but it was many years before engineers broke away from the tradition of direct current that was established at this time.

Another experimental generator was built in 1833 by an American, Joseph Saxton. In his design a pair of bobbin-wound coils rotated just beyond the poles of a stationary horseshoe magnet. In the next few years similar machines, still very small, were developed in response to a prevalent medical fad. It was widely believed that a weak electric cur-

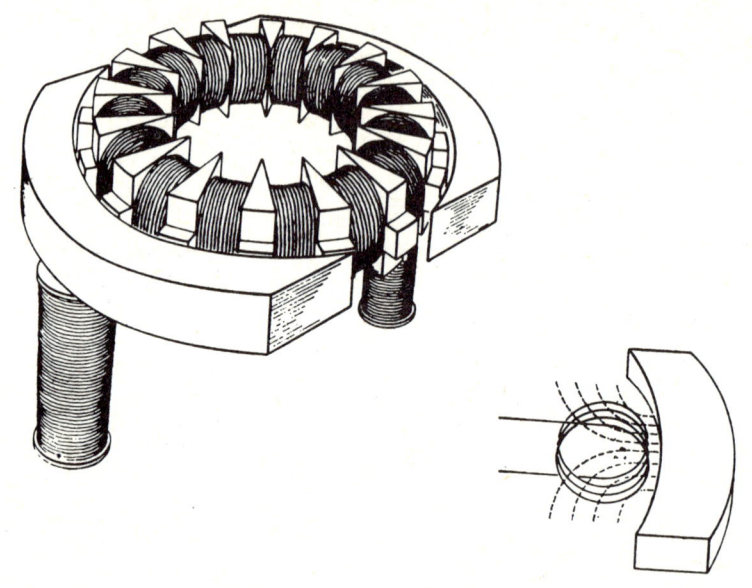

II-7. *PACINOTTI'S GENERATOR of 1860 introduced the ring-wound armature, improved the magnetic circuit and utilized electromagnets. The advantage of his design (illustrated at right) was that the iron ring provided a good path for the magnetic flux and more of each coil was in position to cut lines of force at right angles. But the inner portion of each turn of wire was still ineffective.*

rent sent through the body had a therapeutic effect. This was the first "commercial" application of Faraday's discovery.

The first real economic impetus, however, appeared only in 1839 with the invention of electrotyping. This process, in which copper is deposited by electrolysis on a mold of an engraving, shortly led to the general development of electroplating. Now there was a need for more current than batteries could conveniently supply.

An English engineer, John Stephen Woolrich, saw the possibilities here for an electric generator, and in 1842 he patented a modification of the spinning-bobbin machine. He increased the strength of the magnetic field by stacking several flat horseshoe magnets together and improved the design

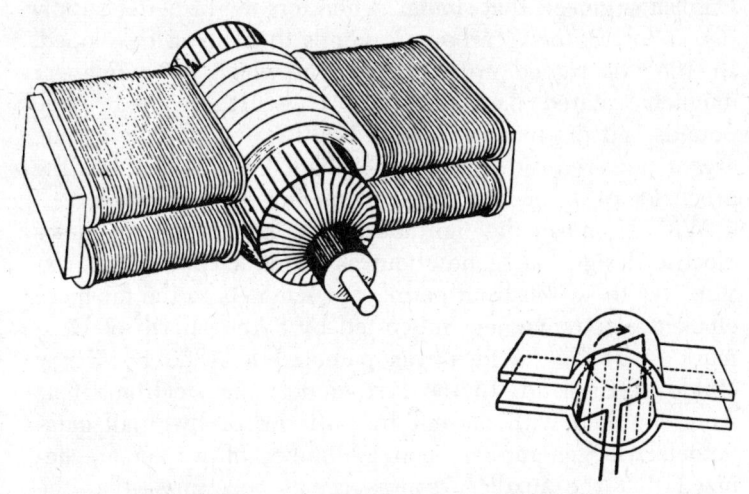

II-8. *HEFNER-ALTENECK'S GENERATOR of 1872* brought the final basic step in armature design: the drum winding. In this method (as shown at right) almost all the armature wire was arranged so as to cut perpendicularly through the lines of magnetic force.

of the commutator to produce the more constant current required for electroplating. Woolrich's generator was driven by steam instead of by hand, and it delivered useful amounts of current for an industrial process, but it was essentially a beefed-up version of the simple permanent-magnet machine, or magnetoelectric generator.

About a decade later a much broader field of application began to open—electric lighting, originally for lighthouses. A Frenchman, F. Nollet, seems to have been the first to think of the dynamo in this connection, but he did so in a rather indirect way. His source of illumination was to be a "limelight," a block of lime heated to incandescence by an oxyhydrogen flame. He proposed to get the necessary oxygen and hydrogen from electrolysis of water, and for this purpose constructed a generator in which a number of coils rotated past the poles of horseshoe magnets. The system was unsuccessful but it suggested to Frederick Hale Holmes, an

English engineer, that similar generators might better supply power for the new carbon-arc lamps then being developed. In 1857 he rigged up a machine in which 36 permanent magnets rotated past stationary coils. It weighed 4,000 pounds and produced less than 1,500 watts. But the carbon arcs it powered did provide a brilliant light, and it was the precursor of several practical, if inefficient, generators.

With Holmes's big lighthouse installations the magnetoelectric design had gone about as far as it could. The next step was to switch from permanent magnets to the far more efficient electromagnet. Suggested by Søren Hjorth of Denmark in 1855, the idea was patented in 1863 by Henry Wilde of England. In the first models the electromagnets were supplied with current by batteries or by small magnetoelectric generators. Soon a number of workers recognized that the auxiliary source is not necessary—that the generator itself can supply the current needed to excite the magnets. The small amount of residual magnetism that always remains in the iron core of the electromagnet provides the initial field, and thereafter the strength of the field increases as the output of the generator builds up.

The self-excited "dynamoelectric machines" represented a considerable advance over the earlier magnetoelectric machines. In their armature design and magnetic circuits, however, they still reflected serious gaps in the theoretical understanding of induction. In 1860 an Italian physicist, Antonio Pacinotti, built a machine that incorporated large improvements in both. First of all he wound his coils on a ring that revolved in the plane of the lines of force between two electromagnets. This arrangement put a larger proportion of the winding in position to cut the lines of force perpendicularly than could any type of bobbin armature. Secondly, he made the ring out of iron, which, as has already been mentioned, increases the magnetic flux that threads the coil. Pacinotti's description of his apparatus in an Italian scientific journal attracted little attention. The electroplaters, who were still the major consumers of electric current, probably never saw the report, and the physicists who did read it presumably were not interested in application. In this case a gap between discovery and application resulted from lack of communication between scientists and engineers.

The ring winding was rediscovered by Zénobe Théophile Gramme of France in 1870. In principle his machine did not differ at all from Pacinotti's, but Gramme was associated with capable businessmen, who saw to it that the invention did not go unnoticed. Very soon the Gramme machine and modification of it became standard equipment both for electroplating and for arc lamps in lighthouses and factories.

The final improvement in armature design came in 1872 with the invention, by F. von Hefner-Alteneck in Germany, of the drum winding. On a ring only the outer portion of each turn of wire produces a useful voltage; the voltage in the inner portion actually works in the wrong direction. The drum eliminates the inner portion completely and puts a much greater length of wire in a position to cut the field perpendicularly.

Hefner-Alteneck had started with a wooden drum but he shifted to iron. At that point the generator had almost reached its present form. Subsequent builders discovered the importance of minimizing the air gap in the magnetic circuit. This they did by such measures as curving the pole pieces of the magnets to fit around the armature and countersinking the windings in slots in the drum so that its iron surface could be brought closer to the magnets. In 1886 the British engineers John and Edward Hopkinson showed how to predict the performance of magnetic circuits, thereby finally taking generator design out of the trial-and-error stage.

By 1890 a flourishing electroplating industry, as well as the mushrooming lighting companies, could obtain direct-current generators about as efficient as those available today. One more giant step in electrical technology remained: the generator had yet to be teamed with the electric motor.

When Faraday sought to produce electricity from magnetism he was looking for the reverse effect of the motor principle—the force exerted by a magnet on a wire carrying a current—that he had demonstrated in 1825. But the electric motor was developed along different lines, and for the most part by different inventors, from the generator. Only gradually did it appear that the motor is the simple converse of the generator and the generator's natural complement in industry and transportation. The importance of reversibility was overlooked until central-station plants for lighting demonstrated that electricity is above all an efficient means of transmitting energy over long distances. By the end of the nineteenth century centrally generated electricity was beginning to replace steam as the motive power in railroading and, during the opening years of the twentieth century, in industry generally.

> Although forms of power other than the internal combustion engine have probably had greater influence on the growth of civilization, the fact seems far from apparent on a superhighway during a summer weekend. Particularly in the United States, and to a lesser extent in other Western nations, it has changed a sedentary population to one in constant motion. It has stimulated the building of highways and airports, all but eliminated the horse in agriculture, mechanized warfare, turned petroleum into a golden fluid, and fostered a service industry which employs millions. Its effect on our cultural patterns has been profound.
>
> The internal combustion engine has so much in common with the explosive weapon using gunpowder and the reciprocating steam engine that it is possible to describe its technical aspects in brief space. The mass production techniques which have been evolved for its manufacture are discussed in Samuel Lilley's article on "Automation." And the deeper implications of its widespread use will no doubt continue to engage the attention of sociologists, economists, and religious leaders.

THE INTERNAL COMBUSTION ENGINE

T. K. DERRY AND TREVOR I. WILLIAMS

ALTHOUGH the complexity of the internal combustion engine and the course of its evolution from a primitive to a practical form make it very comparable with the steam engine, its story can be relatively briefly told in the present context. Although, as we have seen, the steam engine neither started the industrial revolution nor even played the decisive part in its early phases, it subsequently had a decisive effect on many technological developments in the nineteenth century. By contrast, although the main features of the modern internal combustion engine had appeared by 1900, its enormous influence on world civilization was not felt until well into the twentieth century. Moreover, as the steam and the internal combustion engines have a good deal in common—the heart of both being a piston in a cylinder—some of the general principles of operation have already been discussed. In passing, it may be remarked that this similarity to the steam engine was to a certain extent a hindrance to the development of the new engine: early inventors tended to adopt features of steam practice that were not suitable for internal combustion engines. An interesting, if extreme, example of the lingering influence of old on new is provided by a French patent for a petrol horse taken out as late as 1897.

Although the internal combustion engine effectively came into the field a century later than the steam engine, its history is just as long, for both can be considered to have derived from the experiments of Huygens and Papin with a gunpowder engine. As will be recalled, the hazards and difficulties of recharging such an engine, and of sweeping out the products of combustion, caused Papin to turn his attention to steam, and so to set in train experiments that resulted in the first

Newcomen engine of 1712. With the internal combustion engine, as with so many inventions, the fundamental idea was conceived long before there existed either the means of putting it satisfactorily into practice or a strong incentive to develop them. In the nineteenth century the ready availability of coal gas, and later of very volatile petroleum fractions for which there was no existing demand, caused thoughts to turn again to internal combustion engines; at the same time, the growing understanding and mastery of electricity provided an extremely convenient system of ignition, although not the only one used in early engines.

But the incentive was still not very great. Although by far the most important modern application of the internal combustion engine is in transport by land, water, and air, this application can scarcely have entered the thoughts of early inventors. Apart from the fact that the steam-train and the steamship must have seemed to many to represent near perfection in transport by land and sea, the fuel of the earliest internal combustion engines was coal gas which, depending on connection with, or frequent recharging from, a main supply, offered no obvious possibilities for locomotion. Moreover, the weight-to-power ratio of early engines was very high and their running speeds were low: not until the very end of the century did light, high-speed, petrol engines make their appearance. In consequence, all the early engines were stationary ones, designed for industrial use.

Although the seventeenth-century gunpowder engine can fairly be regarded as the ultimate ancestor of the internal combustion engine, nearly two centuries elapsed before such an engine was successfully developed. A gas engine patent was lodged in Britain in 1794, and thirty years later gas engines were used for pumping, but even the Great Exhibition included only a single example, one introduced by Drake in 1843 in the United States. Not until 1859, when a French Engineer, Étienne Lenoir, built a gas engine that in most respects was designed according to current steam engine practice, was any real success achieved. Up to date in so far as ignition of the explosive mixture of gas and air in the cylinder was effected by an electric spark generated by an induction coil, it lacked, among other essential features, any provision for compressing the mixture before it was fired. Its performance did not stand comparison with contemporary steam engines of the same power, but it was nevertheless a landmark as the first non-steam engine able to work continuously under industrial conditions. Not for nearly twenty years was a fully successful gas engine built: this was N. A. Otto's horizontal engine of 1876 (Figure II-9), of which 50,000, of about 200,000 h.p. in all, were sold in the first seventeen years after its introduction by the German firm of Otto & Langen. This engine worked upon the so-

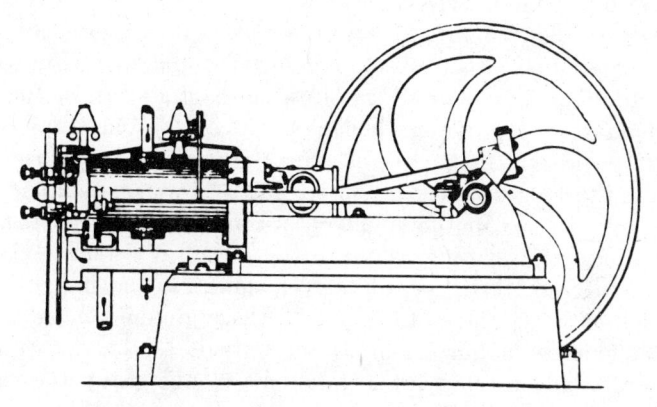

II-9. *Otto horizontal gas engine, c. 1878.*

called Otto cycle, after 1890 almost universally used for all internal combustion engines. The principal exceptions are cheap lower power engines—designed for motor-mowers, lightweight motor-bicycles, and similar purposes—which often work on the simpler two-stroke principle: compression was applied to this type of engine by Sir Dugald Clerk in Scotland in 1878. The Otto cycle, controlled by valves which regulate the intake of fuel and expulsion of the products of combustion, consists of four strokes. In the first stroke the explosive mixture is drawn into the cylinder; in the second, the mixture is compressed by the piston and then ignited; in the third, the force of the explosion drives the piston back again; and in the fourth, the returning piston drives out the gaseous products of combustion, ready for the cycle to be repeated. Although probably an independent invention, the Otto cycle was clearly anticipated in an unexploited French invention of 1862.

Otto's success effectively demonstrated the possible uses of gas engines and by the end of the century their improved size, efficiency, and reliability, and the development of special gaseous fuels, made them fully competitive with steam engines. In 1881 the largest gas engine was of 20 h.p.: by 1917, at which time it reached the peak of its popularity, 5,000 h.p. engines were in use.

Meanwhile, important developments in fuel utilization had taken place that were to determine the main line of evolution of the internal combustion engine: liquid fuels derived from petroleum began to supersede gas derived from coal. For this, two principal reasons are apparent. While the long-distance piping of gas is now commonplace, this was not so in the nineteenth century. Gas was then available only relatively close to the works at which it was manufactured: this generally served domestic and industrial consumers in a single town area. Although the power requirements of nonurban areas throughout the world were growing, only the biggest installations would justify the establishment of gas-manufacturing plant for power alone. To use a liquid fuel, easily transported and stored, capable of being fed to engines by gravity, and yielding more heat per unit of weight than coal, was therefore very attractive. Moreover, at just this time the petroleum industry—originally developed, as we have seen, to meet the demand for illuminants and, after 1878, for heating—could offer just such a fuel at a competitive price. Although the dangerously volatile lighter fractions, which at the outset presented a serious disposal problem to the petroleum industry, were destined to be the most important fuels for internal combustion engines, higher kerosene fractions were used first.

The use of fuels of relatively low volatility presents technical problems different from those involved in the use of petrol, which vaporizes freely at normal temperatures. To prepare a suitable explosive mixture with air in the cylinder, fuel oils must either be vaporized by heating or converted into an exceedingly fine spray. Once an engine is running, the vaporization of the oil and the explosion of a properly balanced mixture in the cylinder of an oil engine can be effected spontaneously by the great heat generated in the compression stroke. While this has the very great advantage of making it possible to dispense with an external ignition system—which even today is the most important single cause of failure of petrol engines—it has attendant disadvantages. The very high degree of compression necessary for spontaneous ignition demands stronger, and therefore heavier, construction, so that oil engines do not lend themselves to lightweight, low power design; moreover, they tend to run roughly at low speeds. In consequence, oil engines have found little favor for motorcars, but they are very extensively used for the heavier forms of road transport, and as large stationary and marine engines they have been enormously successful.

Although the ignition of the mixture in the cylinder of an oil engine will eventually take place spontaneously as a result of the heat of compression, some means of starting from cold must normally be provided: only exceptionally high compression ratios will bring about ignition of oil-air mixtures merely by turning the engine over a few times. In an engine built by Brayton, an American engineer, in 1873, initial firing was facilitated by turning the engine over with the aid of a cylinder of compressed air, recharged by the engine itself once spontaneous ignition had been established. A more successful oil-engine, built in sizes up to 100 h.p., was that patented by Dent & Priestman of Hull in 1886. Operating on the four-stroke Otto cycle, it was started by preheating the cylinder; the success of a mobile unit introduced in 1889 for agricultural use is a reminder of the growing power demand of rural areas. In these engines the heavy-oil fuel passed before combustion through a preheater heated by the exhaust gases. A notably successful engine was the Ackroyd-Stuart, first produced in England in 1890 and manufactured in large numbers by Ruston & Hornsby of Lincoln.

The Diesel engine, first patented by Rudolf Diesel in Britain in 1892, and first successfully manufactured in 1897, is notable for the very careful attention given to thermodynamic principles in its design: these principles Diesel set forth in detail in 1893 in his *Theorie und Konstruktion eines Rationellen Wärme-motors*. His object, eventually shown to be not wholly practicable, was to prevent the engine temperature rising above that of the compressed gas in the cylinders, so making cooling unnecessary; he sought also to increase efficiency by lowering the temperature of the exhaust gases. Diesel's thermodynamic ideals were never fully achieved in practice, and after a few years the main, but exceedingly important, distinction of the Diesel engine from other oil engines was its exceptionally high compression ratio, which favors high thermal efficiency.

The essential principles of the petrol engine are the same as those of the gas and oil engines already described: the main differences are in the systems of fuel injection and ignition and in the fact that it is essentially a high-speed engine. Although an Austrian engineer, Siegfried Markus, is reputed to have built in the period of 1864–1874 several vehicles propelled by petrol-engines, the acknowledged pioneer was the German engineer, Gottlieb Daimler, who had for some years been interested in the design and construction of gas engines. His first petrol engine, patented in 1885, was a single-cylinder vertical machine, air cooled, working on the Otto cycle. The explosive mixture was prepared by sucking air through petrol in a float-chamber, and it was ignited by an externally heated tube inserted into the cylinder head. In the following year this engine was successfully applied to a bicycle, then to a carriage. Within three years Daimler had constructed a two-cylinder engine, in which the two pistons drove one crank: originally it was proposed to fit valves in both pistons, but this intention was never realized in practice. This engine was built in considerable numbers, not only for the motor car industry but also for small boats and as a stationary engine.

Simultaneously with Daimler, another German engineer, Karl Benz, was building engines that were specifically for motor cars. His single-cylinder engine of 1885 differed from Daimler's in being horizontal, in having electrical ignition, and in being relatively slow running. Applied first to a three-wheeled car, the engine proved conspicuously successful when it was used, as a 3½ h.p. unit, in a four-wheeled vehicle in 1893; production of this continued up to 1901. Its system of ignition by an electrical induction coil—powered by an accumulator and fitted with a rotary contact-breaker, driven from the engine, to ensure that sparking occurred at the right point in the four-stroke cycle—was soon widely adopted by other manufacturers. The spark itself was produced in a removable plug, of essentially modern design, fitted into the cylinder-head. The carburetor was of the surface type, similar in principle to that of Daimler's engine, except that volatilization was assisted by the heat of the exhaust and there was a shutter that served as a choke to control the admixture of air.

After 1893 Daimler, and subsequently other manufacturers, used the modern float-feed carburetor invented by Wilhelm Maybach. The petrol level in the chamber of this carburetor is kept constant by a float working a needle valve. The float-chamber communicates, through a very fine jet, with the inlet of the cylinder: the suction from the cylinder causes an exceedingly fine spray of petrol to be injected into the air-intake. Up to the First World War the only other alternative to the original surface-type and the float-feed carburetor was the Lanchester wick carburetor fitted to Lanchester engines, which we shall consider later, from 1897. In this, air was drawn over a series of wicks dipping into a small chamber constantly replenished from the main fuel tank.

Within the period now under consideration, only single- or twin-cylindered petrol engines were in use; multiple-cylindered engines came into general use much later. Although, as we have seen, Daimler introduced a twin-cylindered engine in 1889, the design of such engines presents considerable engineering problems. The fundamental difficulty is that with a four-stroke engine there is only one power stroke for every two revolutions of the crankshaft. From the point of view of engine balance it is desirable, supposing that the cylinders are arranged alongside each other, that one piston should be rising as the other falls. The effect of this, however, is that the crankshaft receives power impulses in two successive half turns, but receives no power impulses at all during the next full turn. This can be overcome if the pistons rise and fall together—one making the first stroke of the cycle while the other makes the third—but this throws the engine out of balance and imposes excessive strain. A solution, but one not widely adopted because it has attendant difficulties, is to use horizontally opposed cylinders. Within the period in question, other ingenious devices were used by a number of manufacturers, of whom perhaps the most successful was F. W. Lanchester. His engine applied the principle of horizontally opposed cylinders, but instead of one crankshaft there were two, one above the other. Each piston was connected to both crankshafts, which carried separate flywheels, but which were geared together so that they turned in opposite directions; the chain-drive to the rear axle was, of course, taken from only one crankshaft.

Most early engines were air cooled: Diesel, as we have seen, paid particular attention to keeping the engine temperature low in order, among other things, to obviate the need for a cooling system. The cooling effect of the air can be increased by increasing the external surface area of the cylinder by means of fins, and the draught resulting from the vehicle's own motion also assists, and can if necessary be increased by a fan, as in the Lanchester engine. The circulation of the lubricating oil—effected by hand pumping on many early cars but already mechanized in the first Lanchester—also promotes cooling. Additional cooling by means of a water-circulating system had, however, also appeared before 1900: it was a feature of Henry Ford's first motorcar of 1896 and also of early Lanchesters, though not of the original model.

Such, in brief, was the main line of evolution of the internal combustion engine up to 1900, by which time many tens of thousands had been constructed. Although its full social impact was not felt until the twentieth century, it is to be noted that all the fundamental problems had been solved: it is fair to say that a modern motorcar engine contains no feature that would surprise a nineteenth-century engineer. The greatest change has been from custom building to mass production, and even the latter lies only just beyond our period, for the Ford Motor Company was founded in 1903.

A very significant feature of the evolution of the internal combustion engine in general, and of the petrol engine in particular, is the relatively small contribution made by British engineers; Lanchester is a conspicuous exception. This becomes all the more striking if comparison is made with the history of the steam engine. The way in which Continental and American inventors outnumbered British supports the general thesis that the Great Exhibition of 1851 marked the climax of Britain's inventive supremacy.

In 1919, after many difficulties and delays, a pamphlet entitled "A Method of Reaching Extreme Altitudes" was finally published by the Smithsonian Institution and its author, Dr. Robert Hutchings Goddard, given $5000 to continue his researches in rocketry. The report, which stated that rockets that would hit the moon could be constructed, was considered by most thinking people the work of a madman, and there were outcries against the dangerous experiments he was conducting. But the paper has proved to be an important engineering document of the twentieth century and the work of Goddard has served as a basis for the development of rocket engines and travel in space.

Goddard was born in Worcester, Massachusetts, on October

5, 1882, studied at the Worcester Polytechnic Institute, and later became an instructor at Clark University. He was testing small rockets while still in his twenties, and at thirty made the computations which indicated that reaching extreme altitudes through rocket power was within the realm of practicality. On March 16, 1926, after several previous experiments, Goddard shot off a rocket at Auburn, Massachusetts which traveled 184 feet at an average speed of sixty miles per hour.

Although Goddard's rocket was the first to employ liquid fuel, the use of reaction motors, as G. Edward Pendray points out, is as old as the skyrocket, and the theoretical understanding of them as old as Isaac Newton's laws of motion. It is likely that in the thirteenth century the Chinese used rockets as engines of war. The English engineer, Sir William Congreve, employed them with success against the French and the Danes at the beginning of the nineteenth century, and inspired the phrase "the rocket's red glare" in "The Star Spangled Banner." But he could not compete with the rifled cannon for accuracy and his experiments fell into limbo.

The Russians claim that the original modern pioneer in rocketry was an obscure schoolmaster named Konstantin Tsiolkovsky, whose first article on the subject appeared in 1903. The Germans, with such students as Hermann Oberth, who published a short mathematical analysis of rocketry and space travel in 1923, were also extremely active. Indeed, their experiments were the first to bear practical fruit in the form of the V-2 rockets of World War II.

Since World War II the development of reaction motors has been staggering. An attempt to land men on the moon will almost certainly be made during the 1960's. Any article on contemporary aspects of the subject is outdated almost before it appears. For this reason, one of the classic nontechnical studies of reaction motors is here reprinted. While speeds, horsepowers, ranges, and fuels may change, the basic theory remains the same. G. Edward Pendray, who was one of the founders of the American Rocket Society, explains it lucidly.

REACTION MOTORS:
HOW THEY WORK AND SOME EXPERIMENTS WITH THEM

G. EDWARD PENDRAY

THE SKYROCKET, which aside from Heron's toylike contrivance* was the first artificial device to make use of jet propulsion, was invented more than seven hundred years ago. But neither Heron nor the hundreds of generations of fireworks makers . . . had any real understanding of the principle of jet propulsion. They only knew it worked.

It remained for Sir Isaac Newton, some 265 years ago, to give us the basis for understanding what rocket power really is, and the unique things it can do. Newton, formulating in simple language the three Laws of Motion his mathematics and observation had helped him to discover, wrote out in Latin this observation: "To every action there is always an equal and contrary reaction; the mutual actions of any two bodies are always equal and oppositely directed."

Thus, the hand that pushes a cradle is itself pushed *by* the cradle, to exactly the same degree and in the opposite direction. The foot that thrusts downward on the earth is thrust upward by the earth in precisely the same amount. The bullet that is ejected by a gun causes the gun to recoil— and the two actions are not only opposite in direction, but are equal in amount.

This is the statement of Newton's Third Law of Motion. Although it describes a phenomenon we daily experience throughout our lives, few people consider or even recognize the reaction that necessarily is a part of every movement of every object. It is important that we recognize it now, for the Third Law is a complete statement of the principle upon which the reaction motor operates.

*See pp. 14–15—*Eds.*

In most human activities the action is what is wanted; the reaction is thrown away or ignored. In jet propulsion, the "action" is thrown away. The reaction is the particular harvest we are seeking.

The simplest form of reaction motor—and the best known—is the one that drives an ordinary skyrocket.

Here is a cross-section drawing of a skyrocket. At the tip is a cone-shaped cap which provides rudimentary streamlining to aid the rapid upward flight of the projectile. Immediately under the cap usually are nested the combustible pellets, the "stars" that cascade brilliantly into the sky at the top of the flight. These are the payload of the skyrocket; they are not a basic part of the rocket itself.

Into the main body of the rocket, usually contained in a heavy paper tube, a quantity of black powder is packed. This is the fuel or propellant charge, *(A)*. The material is usually a form of ordinary gunpowder, often mixed with extra charcoal or some other material to slow down the rate of combustion. It is squeezed into the rocket under high pressure, thus packed tightly into a solid cake. Because it is solid, the flame cannot permeate the cake, so combustion takes place only at the exposed surface of the cone-shaped blast chamber, *(B)*.

The simple thrust mechanism—or motor—of the rocket is completed by constricting the walls of the case below the blast chamber to form a nozzle. Sometimes the throat of the nozzle is reinforced with clay or other hard material to prevent its burning out. A fuse *(D)* and a long stick—a crude balancing device—complete the rocket.

On firing, what happens is this:

Heat from the fuse ignites the surface of the powder on the walls of the cone-shaped blast chamber. The powder does not explode, but a continuous combustion takes place very rapidly, releasing large quantities of gas at high temperature. Considerable pressure builds up instantly in the chamber, since the hot gas is formed at a much faster rate than it can easily escape through the restriction at the nozzle. The net effect is to eject a stream of gas at great

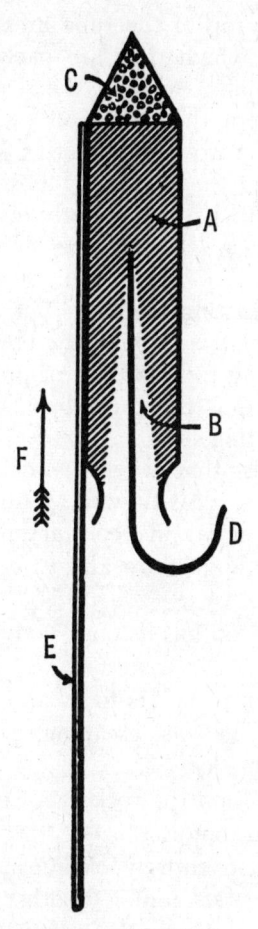

II-10. *Cross-section of a skyrocket.*

velocity, directed backward. This thrusts the rocket forcibly in the opposite direction.

As the fuel burns, the blast chamber rapidly enlarges, but the restriction at the nozzle continues to keep the pressure high and guides the escaping jet. The rocket takes off with a tremendous swish, emitting a stream of sparks and fire, and flies until the fuel is completely consumed. Then an arrangement at the top of the tube fires the "stars" and the bursting charge in which they are packed.

In a jet motor such as that of the skyrocket there are no moving parts—except the stream of escaping gas. It is by no means easy to grasp just how this jet, with nothing to push against, exerts the surprising power that thrusts the whole body of the rocket so violently toward the sky.

The common notion is that the jet does its work by pushing against the air. Superficially this seems reasonable. The air is certainly a resisting medium. But a stream of gas, no matter how rapidly it is moving, or how dense it may be, is no solid connecting rod, capable of pushing against something and transmitting the push back against whatever is adjacent to its starting end.

It is something else that drives the rocket—and this brings us back to Newton and his Law of Motion: "To every action there is always an equal and contrary reaction; the mutual actions of any two bodies are always equal and oppositely directed."

Consider the ejected gas as one "body"; the rocket as the other.

The rocket, forcing the gas to escape, pushes it violently toward the earth. The gas, escaping, pushes the rocket as violently toward the sky.

This is jet propulsion, or rocket power, the simple principle of the reaction motor.

It may still not be entirely clear exactly how this mysterious reaction occurs. Here is another drawing that may help in the further understanding of it.

Imagine a hollow box, filled with gas under pressure. For convenience, let us say the pressure is 100 pounds per square inch. This means, of course, that every square inch on the inside of the box will be pushed upon by gas with a steady pressure of 100 pounds. Under these conditions the box itself will not move in any direction, for the total pressure on each side will exactly balance the total pressure on the opposite side.

Now, suppose a hole were to be cut in side *B*, exactly

Pendray: REACTION MOTORS

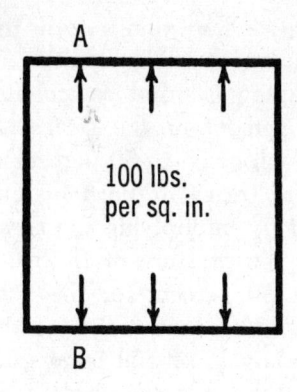

II-11.

(for convenience in doing our mental arithmetic) one square inch in area. At once we have a new condition in the box, as shown in the second diagram. Note that the pressure exerted by the gas on side B is now no longer exactly equal to that on side A. It is, in fact 100 pounds less. The box will therefore be pushed in the direction of the arrow E, with a force of 100 pounds. The push will continue as long as we bring in fresh supplies of gas or fuel, as at D, to keep the pressure in the box at the established level.

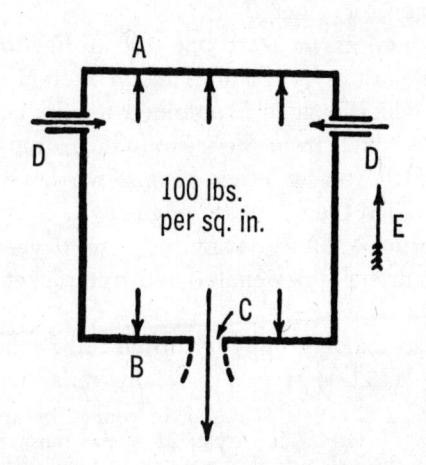

II-12.

In essentials, this is how things go in the reaction motor. An important refinement is the addition of a nozzle at C, to direct the escaping gas and control its expansion. This facilitates the movement of the stream and adds to the thrust, often as much as 35 to 50 percent or more.

The approximate thrust or push of any reaction motor may be calculated by multiplying the area of the orifice of the throat (C) by the pressure of the gas in the chamber, and adding about 50 percent for the additional contribution of the nozzle. The exact value of the thrust will, of course, vary according to several factors, including the fuel used, the pressure of the gas, completeness of combustion and an experimentally determined item rocket engineers call "the constant of the nozzle," which depends upon the nozzle size, shape, and other factors.[1]

So far as the basic principles are concerned, all jet-propulsion motors obey the same laws, and in that sense are identical. When it comes to practical application of these laws, it is soon seen that a bewildering number and variety of motors can be contrived, ranging all the way from low-powered water, steam or compressed-air jets to motors that produce a high-speed blast by burning gasoline, alcohol, benzol, hydrogen, or other fuels.

Moreover, there is at least one broad line of division among all the various types of reaction motors, according to the way in which oxygen is obtained for the combustion. If the oxygen is taken from the surrounding atmosphere, by compression of the air or other means, we have the class appropriately called the airstream engines.[2] If oxygen is supplied in the form of liquid oxygen or some oxygen-yielding compound, the motor is designated as a true rocket or chemical fuel motor.

Each of these major groups is further subdivided. There are, for example, at least two basically different types of

1. The thrust can also be calculated, of course, by applying the formula $T = MV$, where T is thrust, M is the mass ejected per second, and V is the velocity of the jet in feet per second.
2. Also called airjet engines, etc.

the true rocket motors, and at least three types, so far, of the airstream engines.

TYPES AND CLASSES OF REACTION MOTORS

I. The true rocket, or chemical fuel motors:
  A. The solid-fuel rocket motor (burns solid fuels such as gunpowder or smokeless powder).
  B. The liquid-fuel motor (burns fuels such as liquid oxygen and gasoline, or liquid oxygen and alcohol).

II. The airstream engines:
  A. The thermal jet engine (burns a variety of possible fuels, including gasoline and kerosene, with air delivered into the combustion chamber by means of a rotary compressor).
  B. The intermittent duct engine (burns gasoline or similar fuels with air, compressed into the blast chamber by the ram effect. Action is pulsating or intermittent).
  C. The continuous duct engine (burns the same fuels as the intermittent engine; also depends on atmospheric oxygen compressed by ram effect, but its action is continuous).

From this table we see that there is first of all the simple solid-fuel rocket motor, the kind that drives skyrockets and many types of military rockets such as the bazooka, the airplane rockets, and the British "Z-gun" anti-aircraft rocket.

Second, there is the liquid-fuel rocket motor, which provides power for many types of thrusters, catapults, and some types of military rockets, also for such long-range rockets as the German "V-2" rocket weapon.

Among the airstream engines, the best known is the thermal jet engine, also called the "turbo-jet engine," the "turbo-jet" or the "swish." . . .

The other types of airstream engines are duct engines, much simpler than the jet engine in that they require almost no moving parts, and get along without either air compressors or turbine wheels. They do, however, require that the aircraft or glider to which they are attached be given a rapid preliminary start. Duct engines, having no air compressor of their own, depend at least partly on the compression or "ram effect" of the air during flight to provide them with oxygen to burn their fuels.

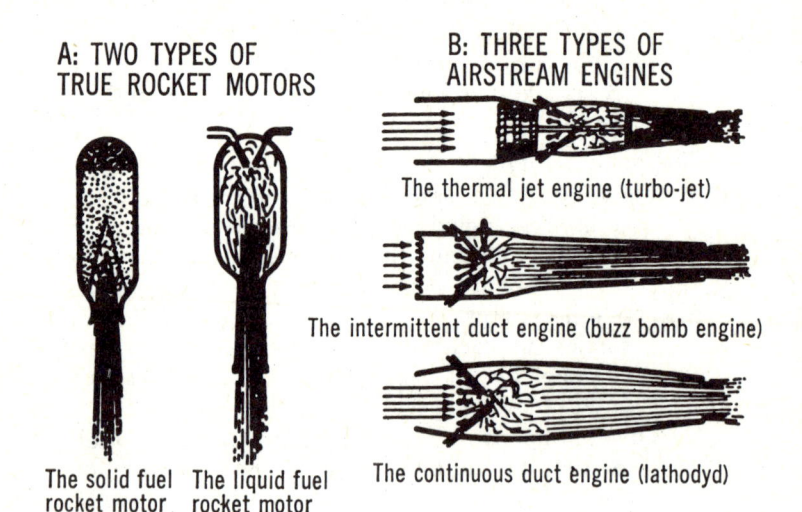

A: TWO TYPES OF TRUE ROCKET MOTORS

The solid fuel rocket motor The liquid fuel rocket motor

B: THREE TYPES OF AIRSTREAM ENGINES

The thermal jet engine (turbo-jet)

The intermittent duct engine (buzz bomb engine)

The continuous duct engine (lathodyd)

II-13. Schematic representation of the five types of reaction motors.

The earliest successful duct engine was the "buzz-bomb engine," used by the Germans to propel their robot bombs in 1944 against London and southern England. It does not operate with a continuous blast, as does the thermal jet engine and the generally used types of rocket motors,[3] but proceeds by a series of pulses or intermittent explosions, hence the name.

Still under development is the continuous duct engine, more usually called the ramjet. In principle it is similar to the intermittent type, depending on the ram effect of the air

[3]. It should be noted that the true rocket motors may also operate intermittently. An intermittent type of dry-fuel motor was developed by Dr. Robert H. Goddard as early as 1916.

for compression, but at suitable high velocity it will produce a continuous blast of power.

Historically, of course, the solid-fuel rocket motors were the earliest to appear. If we look once more at the skyrocket, the most familiar of the true rocket motors, we will perceive that the motor and the fuel supply are there most ingeniously combined. The fuel is made to provide its own motor, in the form of the cone-shaped blast cavity which rapidly enlarges as the fuel burns away.

This form of rocket motor offers no problems to metallurgy, because there is no metal. There is no problem about the burning out of the walls, for so long as they consist of fuel, burning is what is desired. This rocket motor obviously does very well the job it is designed to do. Unfortunately it does not provide us with a good pattern for rocket motors of more power and utility.

For one thing, the gas pressure is necessarily low, whereas good efficiency requires a relatively high pressure. Why is the pressure low? Because if it were higher, it would burst the paper tube in which the rocket is contained, or blow out the flimsy nozzle—or more likely still, cause the flame to permeate the powder charge, no matter how tightly packed. This might set the whole fuel supply off at once, resulting in an explosion instead of a flight.

Another problem with the skyrocket motor is that the combustion chamber changes shape and size continually throughout the run. At the beginning it is small. During the burning it rapidly enlarges. But since its enlargement depends both upon the rate and uniformity of the combustion of the fuel, the shape changes, too, along with the size. If for a given fuel, gas pressure, nozzle, and other conditions there is only one best shape, the skyrocket can have it, at most, for only a fraction of a second. During the rest of its operation it is bound to be doing a less satisfactory job.

The first considerable improvement was the application of a metal nozzle, properly shaped to give maximum aid to the escaping gases. Such a nozzle immediately improves the performance of the motor. Pressure can then be increased by substituting smokeless powder or cordite for the fuel, and exchanging metal or strong laminated plastic containers for the paper ones. This, however, may introduce a new problem: the new containers may weigh more than the paper, and if the chamber pressure is to be increased a great deal, we must be careful to see that the extra weight of the stronger jacket does not nullify the gain obtained from a more powerful fuel.

Dr. Goddard, whose pioneer work established the modern period of rocket research, gave considerable study at one time to the problem of improving the performance of the solid-fuel motor. He discovered that the average velocity of ejection in an ordinary skyrocket was only about 1,000 feet a second, but when he fired charges of dense smokeless powder in strong steel chambers, with properly shaped, smooth tapered nozzles, he obtained velocities of ejection up to nearly 8,000 feet per second. Assuming the masses ejected to be the same, the impulse of the later rocket motor would be eight times as great, with the same weight of fuel, and thus would theoretically drive a rocket not eight but sixty-four times as far—a very considerable reward for thus increasing the jet velocity.

However, it is not possible to add a steel jacket and metal nozzle without adding weight; so as a practical matter the skyrocket and the smokeless powder rocket could hardly start with weights, fuel charges and other factors even. The smokeless powder rocket would have to weigh more, or else carry less fuel or payload at the start.

To get around this it was early suggested by Dr. Goddard that the blast chamber should not be merely a cavity in the fuel supply, but a separate contrivance, into which the fuel could be inserted as needed. Several ingenious ways were suggested for doing this. Dr. Goddard's proposal was to shoot pellets of fuel into the chamber intermittently, like machine-gun bullets.[4] Similar ideas offered by other experimenters include thrusting a solid stick of fuel rapidly into the chamber

---

4. Dr. Goddard patented such an intermittent rocket apparatus in 1914, and brought it to a good state of development during the First World War. He later gave up this line of experiment, however, in favor of liquid-fuel motors.

through an orifice, the speed of insertion being equal to the rate of burning, or powdering the fuel and blowing it in by air or gas, through suitable inlet ports.

If one or another of these ideas were adopted, only the motor chamber (which could be relatively small) would need to be strong enough to withstand the high gas pressure of the blast; the fuel container could be light and flimsy, and thus add very little to the weight.

Unfortunately, any apparatus for shooting pellets into the chamber is likely to be heavy, expensive and cantankerous, full of personal little kinks and problems of its own. Blowing the powder into the chamber with gas pressure is no easy one to handle, either. Likewise, the idea of thrusting a stick of fuel into the chamber through some sort of opening runs into very special headaches, including the difficulty of sealing the edges of the orifice against back pressure without also making it too hard to push the fuel in. Too, there is the matter of judging to a high degree of accuracy just how fast the fuel will burn—solid fuels being particularly variable in this respect.

The upshot is that while the application of one or another of these ideas might possibly further improve the performance of solid-fuel rockets, experimenters and military experts have preferred to continue using solid-fuel rockets of the simpler sort—taking the disadvantages in exchange for the convenience and general freedom from worry that solid-fuel rockets without internal mechanism can have at the site of battle.

This should not be taken to mean that great advances have not been made, however. The demand for powerful, simple, dependable war rockets of many kinds has put great pressure on technical rocket men. The improvements have been almost countless, and include better metal and plastic jackets, proper metal nozzles—and above all, better fuels.

Practically all of the new fuels are related to cordite, of course.[5] There are now not only many types of these pro-

---

5. This statement is no longer true. By 1958 solid fuels of many different types and bases were being employed, and in many instances solid-fuel rockets were competing actively with their liquid-fuel counterparts for long-range-missile and space-flight uses.

pellants, but the charges are being made in a variety of ways, one of the most interesting being extrusion. In this process the material is made in plastic form and pushed through dies—usually in factories operated by remote control—to produce long rods or tubes of the fuel material of just the right size and shape to fit into the rocket bodies. These need only cutting to proper length and final insertion.

In most modern military rockets the fuel is burned with great rapidity. In the bazooka, for example, the whole charge goes off in a fraction of a second, while the rocket is traversing the length of the eight-foot launching tube.

The quick-burning effect is produced by a process just the reverse of that used in the skyrocket, where the powder is tightly packed to keep the flame from permeating it. In the quick-burning solid-fuel rocket, the charge is specially prepared to encourage the flame to get almost everywhere at once. If gunpowder or smokeless powder is used, the load may be in the form of "doughnuts" or pressed fuel, packed with loose powder to propagate the flame. If it is cordite or some other of the more powerful explosives, the charge may be fluted, hollowed, or drilled full of holes, so the flame can eat it quickly, over a large surface.

The effect of this quick burning is to produce a takeoff almost like a cannon shot. The rocket gets away with a quick "ffff-tt." The flame is hardly more than a brilliant flash and then gone. The projectile flies almost all of the way to its target on momentum, obeying the same laws of ballistics as an artillery shell.

For such short-range devices as military rockets, the dry-fuel motor does very well; in fact it is the only practical type. But this kind of motor will never give the power and sustained performance needed for high-altitude sounding rockets, for example, or long-range military or trajectory rockets.[6] For these, we must turn to the liquid-fuel motor.

The solid-fuel rocket has obvious limitations which the liquid-fuel motor appears to overcome readily—but in so doing it introduces a host of new problems all its own.

6. See footnote 5. Solid fuels have undergone very considerable development since this statement was first written in 1944.

To begin with, it is a device that functions only in the presence of intense heat. The temperature within a liquid-fuel rocket motor is almost always at or above the melting point of the materials of which it is constructed. Moreover, there is an enormous contrast in temperatures from one part of the motor to another. At the point where the fuel enters, the thermometer may register as low as the boiling point of liquid oxygen, −297 degrees Fahrenheit. At the hottest point, the temperature may be at least half that of the surface of the sun.

The burden which these conditions put upon a simple, small and necessarily light structure is enormous. The surprising thing is that a liquid-fuel motor made of metals can operate at all. Yet motors have been developed in this country to give sustained performance for indefinite periods, to provide thrusts from a few pounds to 6,000 pounds or more[7] and weighing, for the largest sizes, about 200 pounds, or about ½ ounce per pound of thrust. Smaller ones, capable of yielding thrusts up to 100 pounds, may be nested in the hand, and weigh less than a pound.

The high temperature at which the liquid-fuel motor must be operated to produce suitable results is no mere temporary obstacle which some ingenious trick or discovery may some day solve. Rather, it is inherent in the nature of fuels and jet velocities. It is the principal limiting factor on the operation of any jet motor.

To obtain a jet velocity of between 6,000 and 7,000 feet per second with fuels of the type in general use, the temperature inside the combustion chamber must be around 5,000 degrees Fahrenheit. To obtain a jet velocity of 8,000 feet per second, a blast-chamber temperature of about 8,000 degrees Fahrenheit is required. A jet velocity of 10,000 feet per second will need a temperature of 12,200 degrees, and a 12,000 foot-per-second jet can be generated only by operating at a temperature of some 17,300 degrees. No ordinary con-

[7]. The thrust developed by the German "V-2" rocket motor, a liquid-fuel type, is more than 26 tons. Motors used in long-range missiles and for space flight may have thrusts of one million pounds or more.

structional metals, of course, will stand any of these temperatures, except for very brief periods.

The melting point of aluminum, the most commonly used material for rockets because of its lightness, is 1218 degrees Fahrenheit, or only about a fourth of the temperature required to produce a 6,000 to 7,000 foot-per-second jet. The melting point of steel, which is often used in motor construction, is around 2,200 degrees; still well below the temperature of the 6,000 to 7,000 foot motor.

Almost every rocket experimenter has seen motors of the finest steel burn out in less than a second of firing. At these temperatures it is not so much a matter of melting as of erosion. The metal behaves somewhat like an icicle in the flame of a blowtorch. The surface melts or softens. Then the furious blast of the escaping jet carries the softened part away, exposing new material underneath. Almost before the metal of the motor has become hot clear through, it has been cut to pieces by this process. Under such circumstances all ordinary ideas of cooling are futile.

Dr. Robert H. Goddard was the first experimenter on record to tackle the liquid-fuel motor problem, and was also the first to shoot a rocket powered by a liquid-fuel motor. The German experimenters of the *Verein für Raumschiffahrt* (German Rocket Society), however, were quicker than Goddard to report their experiences. Consequently, the history of their early attempts to harness liquid fuels is a well-known and highly instructive tale.

Their first idea was to place the blast chamber of the motor directly in the liquid oxygen tank, thus allowing the cold oxygen itself to cool the metal. It was an ingenious scheme, but in the initial design the intense heat of the motor raised such sudden pressure that the oxygen tank ultimately exploded. The motor was cone-shaped; a direct descendant of the cavity in the skyrocket's fuel load. The nozzle, which projected out of the bottom of the oxygen tank, was about three inches long. Blast chamber, nozzle, and oxygen tank were made of an aluminum alloy, chosen for its lightness and the rapidity with which it conducted heat.

In their next attempt, the experimenters produced an

oxygen tank which had a larger safety valve, and the motor also was larger. Its performance, however, was very much like the first. In both of these motors the fuels were introduced through separate inlet ports near the throat of the nozzle, and were forcibly injected upward by gas pressure.

The tests taught several lessons, not the least of which was that liquid oxygen is not a very satisfactory coolant. It also became clear that a cone-shaped chamber is not an effective design. The volume of the chamber is too small in relation to the area of its surface, and the sharp corners where the nozzle joins obstruct the rapid flow of the gases.

A later motor developed by these experimenters and demonstrated in 1931 to Mrs. Pendray and me as representatives of the American Rocket Society had many improvements. The motor was cooled by running water. The material was aluminum alloy, and the blast chamber was what the experimenters called "egg shaped," though inspection showed it to be cylindrical, with each end finished off in a hemisphere. The propellants came in through two ports near the throat, directed upward so the streams of liquid oxygen and gasoline would meet near the radius of the upper hemisphere.

This motor was small, weighing about a pound. It produced a thrust of about 20 pounds, at a calculated jet velocity of about 6,000 feet per second.

Perhaps the best known pioneer series of motor experiments ever reported were those made under the auspices of the American Rocket Society between 1932 and the beginning of the Second World War.

The society's first motor was based principally on German designs. I returned to this country with the ideas on which it was constructed following the *Verein für Raumschiffahrt* demonstration in 1931. The motor was not merely a duplication of the German work. Made of cast aluminum alloy, it was considerably heavier than the German motor. The blast chamber was cylindrical, with hemispheric ends, and measured two inches in diameter and three inches in length. The nozzle was three inches long, with a half-inch throat and a taper of about ten degrees.

The inside of the nozzle was carefully machined and the inner surface finished to mirrorlike smoothness. The fuel inlets were bored—after the German fashion—in such a way as to introduce the fuel near the nozzle, in streams directed toward the back of the motor. The theory of this was, simply, that such a position for the inlets would make the fuel travel substantially twice the length of the motor, and thus provide for some cooling effect along the walls as well as better mixing and combustion. It was an excellent theory, but subsequent experiences indicated that it did not work well in practice, and it was later abandoned.

The first test took place at a proving ground near Stockton, New Jersey, on November 12, 1932. The motor was mounted for the test between two pipelike cylindrical tanks, one of which contained liquid oxygen, the other gasoline under nitrogen gas pressure of 300 pounds per square inch. A water jacket surrounded the motor for cooling. The whole contrivance was fastened between two parallel upright wooden bars, on which it was free to slide against the tension of a spring. Previously the spring had been calibrated so that by measuring the distance traveled against its tension, the experimenters would be able to determine the thrust of the motor.

The report[8] on the first test, as published in *Astronautics* for November 1932, read in part as follows:

> We had previously decided that the fuels should be turned on almost simultaneously, the oxygen first, the gasoline close behind. I judged that the fuse was going properly to light the fuels. About three minutes had passed since the final turning down of the oxygen valve. Enough pressure should have been built up to start the firing. . . .
>
> Mr. Pierce (H. F. Pierce, later president of the society) threw his switches rapidly. The fuse apparatus worked to perfection. For an instant there was a great fire, as the pure oxygen struck the burning fuse.
>
> In an instant the gasoline was also pouring into the

8. "The History of the First A.I.S. Rocket."

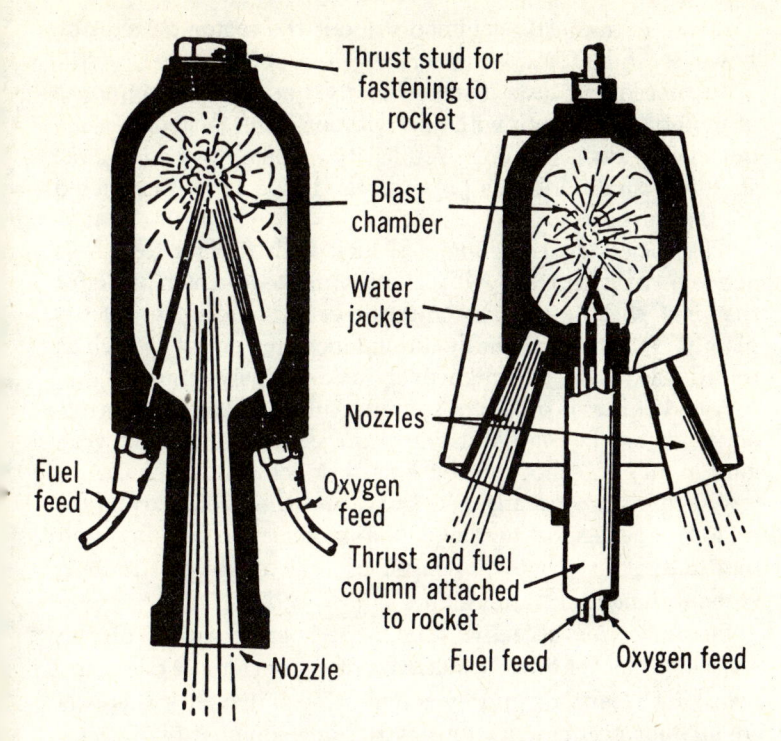

II-14. *Two early types of liquid-fuel rocket motors. Left— the original ARS motor, designed by the author and H. F. Pierce; right, a four-nozzle motor designed by John Shesta for the ARS No. 4 rocket.*

rocket. The fuse, the flare, and the uncertainty about the performance of our rocket motor all disappeared at once, as, with a furious hissing roar, a bluish white sword of flame shot downward from the nozzle of the combustion chamber, and the rocket lunged upward against the retaining spring. . . .

But most important—the marks made by the rocket on the soaped guides indicated that it had registered a lift of sixty pounds. . . . Our fifteen-pound rocket would, in a vacuum, have ascended to a height of sixteen miles. Discounting liberally for air resistance, a well-designed rocket, flying perfectly straight, ought with so much power to reach an altitude in air of five to eight miles.

These enthusiastic comments about the motor turned out, however, to be too optimistic.

In subsequent tests it scored badly, and a whole sequence of motors like it, both with and without water jackets, burned out with disheartening regularity when tested against standard conditions on a proving stand especially constructed for the purpose.

The society had phenomenal luck with its first test. It is not known to this day why the motor stood up so well for that first shot or so, and later showed its weakness all too plainly. It was that kindly Providence, no doubt, which is traditionally said to watch over rocket experimenters.

A rocket motor produces so much quick power, and speeds on its way so fast in flight that it does not have to burn very long in any one shot to go a long distance. But more than a few seconds are usually necessary, and the need, except for military rockets, is to develop a motor that can be used repeatedly, for whatever length of time might be desired, without danger of burning out.

Pursuing this objective, the experimental group of the American Rocket Society subsequently gave almost five years (working weekends and evenings principally, since this was an amateur occupation) toward the development of successful motors capable of firing for indefinite periods, and developing high thermal efficiency.

The first step was the construction of a suitable proving stand. To provide a cheap and easy method of getting data on many types and shapes of motors without building completely new models each time, a sectional motor was developed which made it possible to exchange some parts—for example nozzles—without any more trouble than the undoing of a few clamps.

With this proving stand, followed subsequently by bigger ones and additional refinements, it was possible to try a great many types of motors and motor parts. It was possible, too, to prove to the satisfaction of all concerned that simple cooling schemes, such as water jackets, ice bags, dry-ice packs and the like were quite ineffective if the shot were to last

Pendray: REACTION MOTORS 123

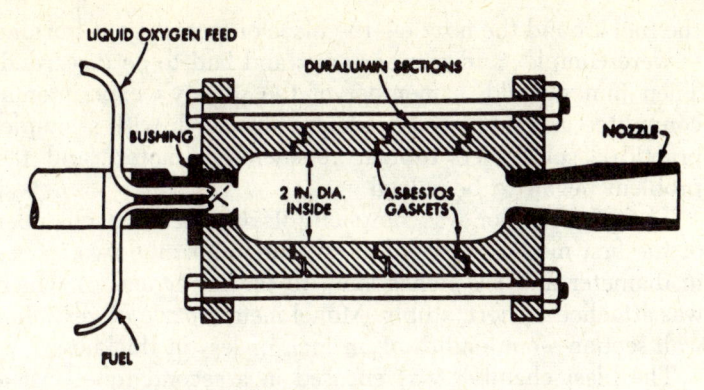

II-15. *Sectional liquid-fuel motor used in the American Rocket Society's test stand experiments.*

more than a few seconds. Air-cooling likewise proved ineffective.

Turning next to metals and materials which it was believed would withstand the heat without special cooling, the experimenters performed tests with ceramics and fire clays of various types, with hard metals such as Stellite, with Nichrome and other heat-resistant metals, and with blocks of pure carbon serving as nozzles. The ceramics and fire clays cracked rapidly under the change in temperatures in the motor. The heat-resistant metals soon proved to lack sufficient heat resistance. The carbon had too little strength; motors made of it burst promptly. All other materials likewise failed with disheartening certainty.

By the end of the first series of tests, it was clear that none of these schemes would produce a permanent non-melting motor. The experimenters next turned to studies of ways in which the incoming fuel or the liquid oxygen could be made to do the cooling. The experiences of the Germans with liquid oxygen as a coolant brought this suggestion into early disrepute. Using the oxygen in this way simply caused it to boil furiously and vaporize before it could get into the blast chamber.

The fuel, however, offered a source of cooling that had possibilities. Most of the suggestions for ways to introduce the fuel around the nozzle—the place of most serious burning—were complex and cumbersome, and had to be discarded. Then James Wyld, a member of the society's experimental committee and later its president, came forth with a simple, practical "self-cooled tubular regenerative motor" and the problem began to be solved.

The Wyld motor was simplicity itself. The blast chamber of the first model consisted of an aluminum tube two inches in diameter and six inches long, to the lower end of which was attached a short, stubby Monel metal nozzle of very thin wall section—one eighth of an inch or less in thickness.

The blast chamber was encased in a second tube, just a trifle larger, so as to leave a cylindrical space about an eighth of an inch in thickness for the passage of the fuel. This double jacketing was carried to the nozzle also, but here the permitted thickness of the liquid layer was greater.

In such a motor the fuel comes in near the tip of the nozzle and goes into the coolant chamber surrounding the nozzle with a swirling motion. From there it passes rapidly through the passage surrounding the blast chamber, and thence to a mixing device at the head of the motor. From the mixer it is sprayed through a series of inlet ports, intimately mingled with liquid oxygen which is brought in at the motor head.

When the walls of the motor are thin enough, heat imparted by the escaping jet can readily pass through the metal and into the incoming fuel. The motor is called

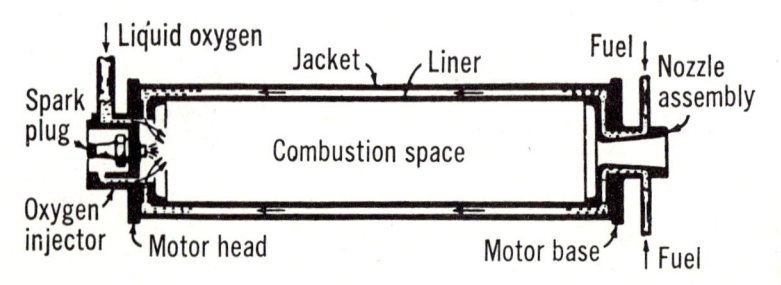

II-16. *Cross section of the Wyld regenerative motor as originally presented in* Astronautics.

"regenerative" because it saves heat that would otherwise be wasted through the nozzle and the blast-chamber body, and brings it back inside the motor. . . .

In basic principle, the Wyld type of regenerative motor is the motor in use today in many an operation making use of liquid fuels. It works almost equally well with liquid oxygen, nitric acid, or other liquid oxidizers. Only the metals used in its construction need be altered to take account of the relative corrosiveness of some of these chemical combinations. To use nitric acid, for example, the inner part of the motor and its connections must be made of stainless steel.

The regenerative motor has a quite pleasing by-product, which arises from its regenerative features, the length of the blast chamber and the general design: its thermal efficiency is excellent.

The efficiencies are now of the order of 40 to 45 percent in larger motors of this type, corresponding to jet velocities of 6,000 to 7,500 feet per second or better with liquid oxygen and gasoline. This efficiency is still far from the theoretical maximum, but compared with the efficiency of the skyrocket, at 2 percent, or the early motors of American experimenters, which were only about 5 percent efficient, this is enormous.

Edition of 1949
Revised by the author, 1958

The search for new sources of energy continues. One of the most promising is the utilization of solar power. The amount of energy reaching the earth in the form of solar rays far outweighs all other kinds of energy now in use, but as yet the problems of harnessing it profitably have eluded man's ingenuity. At present the most stupendous source of useful—or destructive—power is released in atomic fission or fusion. The historical background is brief in span.

The scientific study of atomic disintegration can be traced back only as far as Roentgen's discovery of X-rays in 1895; the equivalence of mass and energy to Einstein's theory of 1905. The feverish activity attendant on experimental disintegration of the atom just before World War II, the engineering miracles accomplished by the Manhattan Project, the bomb which finally brought the Japanese to their knees, and the race for ever more destructive bombs which has taken place since 1945 are parts of one of the most dramatic and important stories in the history of the human race. Less spectacular but, it is to be hoped, of greater eventual importance is the use of atomic energy for peaceful purposes. The British have been among the leaders in such applications, and the former Managing Director of the Industrial Group of the United Kingdom Atomic Energy Authority, who designed and built the first British atomic energy factories, here describes the principles and practices followed.

THE A.B.C. OF ATOMIC ENERGY

SIR CHRISTOPHER HINTON

IN A CONVENTIONAL power station we burn coal or oil to boil water which produces steam: the steam turns the blades of a turbine rather like the wind turns the sails of a windmill: the turbine drives large dynamos which in turn produce electricity. In an atomic power station exactly the same thing happens except that instead of boiling water by burning coal or oil we boil it by using some of the energy locked up inside the atom. This is the only difference. Instead of the conventional furnace in which coal is burnt we have an atomic furnace, called a reactor, in which we generate heat by splitting atoms.

Of all the different types of atom which occur in nature, there is only one with which we can carry out the fission process, and this one is an isotope of uranium. Like all uranium atoms its nucleus contains 92 protons, and it also contains 143 neutrons making a total of 235 particles. So it is called, in scientists' shorthand, uranium 235. This is a very big and complex nucleus and if a stray neutron succeeds in penetrating the barrier of the circling electrons the nucleus is so unstable that it breaks into two pieces—two smaller nuclei. Their combined energy content is much less than the energy content of the original uranium atom, and the surplus energy is given off as heat. This surplus energy drives the fission products apart at enormous speeds. They very quickly lose their speed by colliding with the surrounding atoms, and their energy of movement is converted to heat.

There is one other important effect of this fission reaction. Besides the big pieces, the fission products, separate neutrons shoot out of the splitting nucleus. On the average we get about two and a half neutrons from every fission. Remember we started all this off with one neutron and we now have two or three new ones.

The importance of these new neutrons is that in a lump of uranium 235 they can split still more atoms thereby releasing more heat and producing still more neutrons, and so on in an ever expanding chain reaction. In this way the number of fissions builds up very rapidly and each one produces heat. The whole thing happens so quickly that in a fraction of a second so much heat will be produced that there will be an explosion. In fact, this is what happens in an atomic bomb.

But this fission process—this splitting and release of heat —only takes place with one kind of uranium atom and this kind is comparatively rare. When we extract uranium ore from the earth and purify it, we get a heavy metal which is a mixture of two kinds of uranium atoms, two isotopes, which we call uranium 235 and uranium 238. Uranium 235 we have just considered—it splits when its nucleus is struck by a neutron. Uranium 238 behaves very differently. If a neutron strikes its nucleus the neutron is captured and eventually that atom is converted to an atom of an artificial element called plutonium. Plutonium behaves like uranium 235— when struck by a neutron its nucleus splits, produces fission products and heat, and releases more neutrons. It is, in fact, an artificial nuclear fuel.

We can now see how we should like our nuclear reactions to carry on inside our reactor. Let us suppose we have a lump of natural uranium and we expose it to neutrons—never mind where the neutrons come from at the moment, I shall deal with that later. The reaction starts when one of these neutrons strikes an atom of uranium 235. The atom splits, fission products are formed, heat is given off, and two or three neutrons shoot out. Let us suppose it is two neutrons from each fission, as I did say that on average 2½ neutrons appear but some are inevitably lost. To keep our reaction going one of these two neutrons must hit another atom of uranium 235 but the other can hit an atom of uranium 238. This does not produce heat, but the uranium 238 atom is converted to plutonium which gives us fresh fuel. If we can carry on like this, using one neutron from each fission to cause a further fission, and the other neutron to convert uranium 238 to plutonium, we get a steady production of heat and a steady production of plutonium. The steady flow of heat is really what we are after as it is this heat which we can use to boil water to raise steam in our nuclear power station. The simultaneous production of plutonium is a fortunate bonus or by-product as we call it. But for every atom of uranium that will split and give off heat there are 139 atoms that will not.

We must improve the chances of the neutrons striking the rare atoms of uranium 235, and fortunately there is an easy way of doing this. When the neutrons first shoot out from the splitting atom they are traveling at tremendous speeds, and it has been found that if they can be slowed down sufficiently they are then much more likely to split other atoms of uranium 235 and less likely to be absorbed by the more plentiful atoms of uranium 238. How do we slow them down? We do it by making them bounce about amongst atoms of a substance which is unlikely to capture them. We call the substance a moderator, because it moderates the speed of the neutrons. There are various possible moderators, but the one we have used in most of our reactors in Britain is carbon in the form of very pure graphite. The essential structure of a reactor is a mass of graphite with lumps of uranium spaced regularly in it. In fact the first reactor ever built in 1942 was just that. If the reactor is going to produce heat, we must also provide some means of collecting the heat. We can do this by forcing a gas through the graphite and uranium structure, where it picks up heat from the hot lumps of uranium and we can then use this heated gas to boil water.

Let us now consider what happens to the neutrons. We start with two and a half from each fission, and one of these must cause a further fission if the chain reaction is to continue. Of the rest, some will be captured by atoms of uranium 238, and others by the materials of the reactor. Some will escape altogether. If the graphite and uranium structure, which we call the core, is small, so many neutrons will escape that there will not be enough left to keep the chain reaction going. As the size of the core increases the proportion of neutrons which escapes gets less. When this proportion is low enough to leave just enough neutrons to carry on the chain reaction, we say the core has reached the critical size. Anything smaller, and the reactor will not work at all. Anything bigger will leave us with neutrons to spare.

A practical reactor must be bigger than the critical size so that there are spare neutrons to enable the fission reaction to build up. But we must not let it build up too far, or the release of heat will be too great—our reactor will get too hot. So we must have some means of controlling the reaction, and we do this by absorbing the spare neutrons when the number of fissions has reached the level we want.

Certain substances absorb neutrons very readily, and one of these is boron. Boron can be made into an alloy with steel and if we push rods of boron steel into the core the rods will absorb some of the neutrons and prevent them from splitting uranium 235 atoms. The further we push them in, the more neutrons they will absorb, and in this way we can control the level of the reaction, just as we can control the rate at which an ordinary fire burns by pushing a damper into the chimney. If we push the rods in far enough, we can stop the reaction altogether.

It is always wise to have a reliable safety device, like a fuse in an electric circuit, to prevent an accident if something goes wrong. In the same way we want to be sure of being able to shut down our reactor in case of need, even if something goes wrong with the control rods. We achieve this by having a similar set of boron steel rods hung above the reactor core, and in an emergency they will fall into the reactor under their own weight and shut it down. One final problem has to be faced. If our reactor is producing heat it will inevitably produce fission products and these are highly radioactive. Something must therefore be done to protect the men who operate the reactor from too much exposure to this radioactivity. This we do by building a concrete box several feet thick round the reactor. The concrete will absorb sufficient radioactivity to make it biologically safe, and for this reason we call it a biological shield.

II-17.

Let us now summarize the basic features of a reactor (Figure II-17). On the outside we have the massive concrete box to absorb the radioactivity. Inside this concrete box is our graphite moderator which slows down the neutrons. At regular intervals in the graphite are the lumps of uranium, our nuclear fuel. Now for a variety of reasons you cannot put lumps of uranium into a reactor like you can throw lumps of coal on a fire. You have to make the uranium up into bars and seal the bars in metal containers. This assembly we call the fuel element. As the atoms split in the fuel element the container gets very hot. This is the heat we are after and we collect it by pumping gas past the hot containers and then use the heated gas to boil water. Finally we have our control and safety rods, the control rods to absorb any excess neutrons and so regulate our atomic furnace, and the safety rods to shut it down completely if the need arises.

Now to deal with a point I mentioned earlier. Where do the neutrons come from to start the whole reaction? Well, they could come from anywhere. Wherever you may be at this moment there are a few stray neutrons flying about you. We could rely on such stray neutrons to start off our reactor, but in addition to these we sometimes put inside the reactor what we call a neutron source, which will give us a steady supply.

The first large reactors were at Windscale. These were designed to produce plutonium for defense purposes, but of course they produce heat as well, because as we have seen every reactor of this type inevitably produces both heat and plutonium. When we built them around 1949 we did not know enough to run the reactors at a high enough temperature to be able to harness the heat economically, so that what heat was produced just went up the chimney.

By 1953, when it was decided to build more reactors, this time at Calder Hall, the problems had to some extent been solved and besides using the reactors to make plutonium we were able to use the heat from them to boil water. The steam thus produced drives turbines and the turbines drive alternators, which generate electricity (Figure II-18). Let me remind you of what I said right at the beginning—the only difference between a nuclear power station and a conventional power station burning coal or oil lies in the source of the heat. The rest—the boilers, the turbines, and the alternators—are all the same.

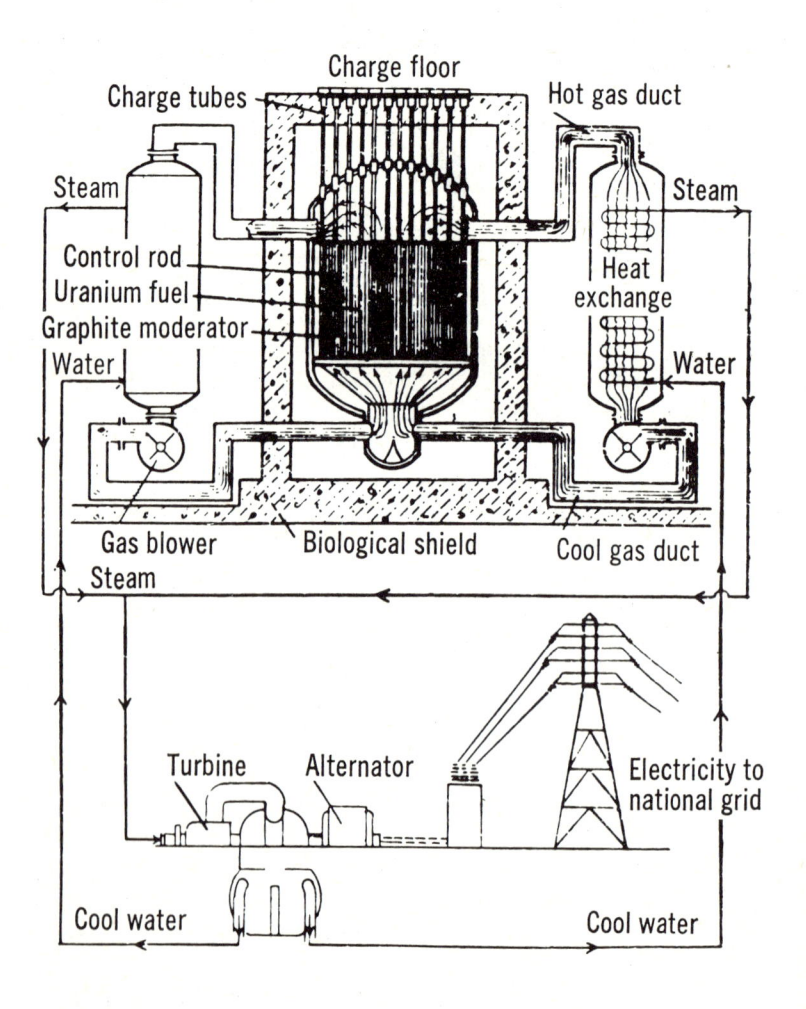

As Calder Hall is the prototype of the civil atomic power stations now being built it is worthwhile to study it in some detail. This will show us how practical engineering requirements led to modifications of the basic design and will also give us some idea of the size and complexity of an industrial reactor.

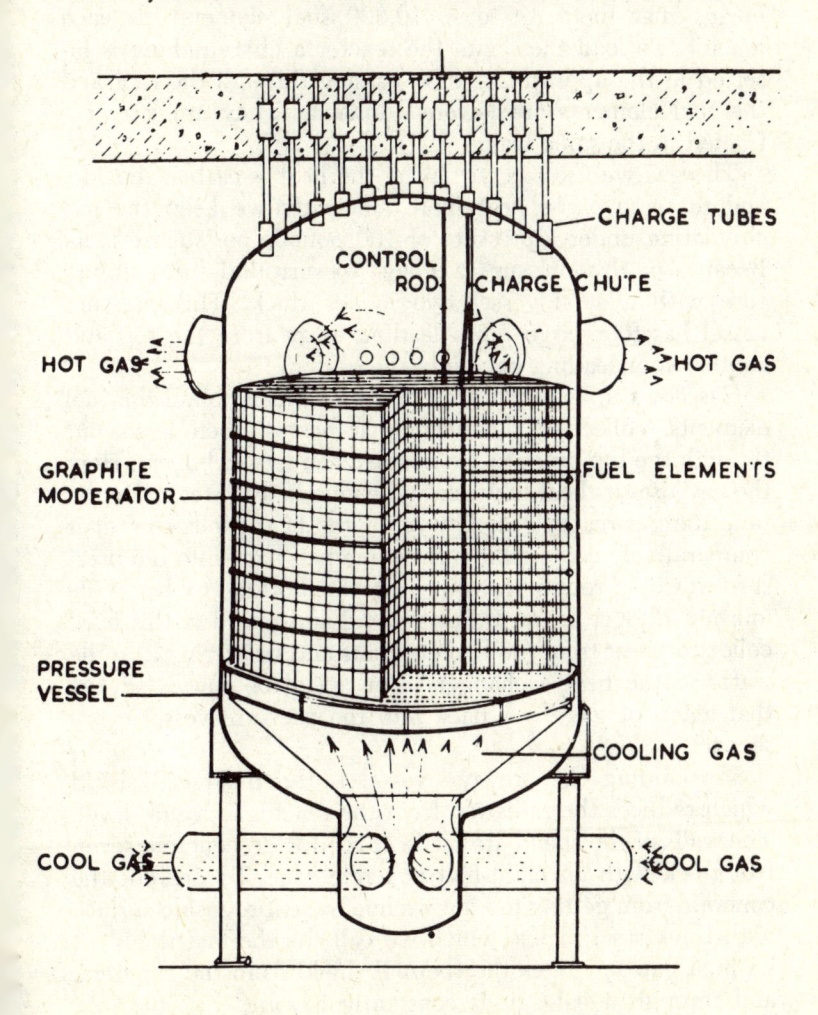

At Calder Hall (Figure II-19) the moderator is made up from 58,000 separate pieces of graphite into a structure 36 feet across and 27 feet high. Every piece was made to an accuracy of two thousandths of an inch. The channels in which the uranium fuel elements are placed run from top to bottom instead of from side to side and there are some 1,700 of them. Each fuel element consists of a rod of uranium about an inch in diameter and 40 inches long and is sealed in a metal container with spiral fins to help in transferring the heat from the uranium to the gas. Six of these fuel elements are stacked one on top of another in each channel, which means that there are over 10,000 fuel elements in each reactor. To load them into the reactor a charge chute is inserted in the appropriate tube passing through the roof and the fuel elements are lowered down it. They are later extracted in the same way.

The gas we use to carry away the heat is carbon dioxide, and to remove the heat more efficiently we keep the gas circulating under a pressure of 100 pounds per square inch. Because of this pressure we have to surround the graphite core with a steel vessel two inches thick. This pressure vessel has four large pipes leading away from the top and another four leading back into the bottom.

Gas flows up through each of the channels round the fuel elements, collecting the heat as it passes, then flows out through the top pipes to the heat exchangers or boilers. Here the gas loses its heat to water and converts it to steam. By the time the gas reaches the bottom of the heat exchangers it is comparatively cool and it is then pumped back into the pressure vessel through the pipes at the bottom to begin its journey all over again. It flows round and round in this way, collecting heat from the fuel elements and delivering it to the water in the heat exchangers. The speed of flow is so fast that a ton of gas flows back into the pressure vessel every second.

Surrounding the pressure vessel is the biological shield which reduces the radiation from the reactor to a safe level. The walls of the shield are made of reinforced concrete seven feet thick with an eight-foot concrete roof. To prevent the concrete from getting too hot we line the entire inside surface with thick steel plates, which we call the thermal shield. A six-inch gap separates the thermal shield from the concrete, and through it cold air is constantly flowing.

The top of the eight-foot concrete roof forms the working deck of the reactor. From here the fuel elements are loaded into the core thirty feet below through tubes which pass through the concrete and into the pressure vessel itself. In order to reduce the number of openings in the pressure vessel, each of these charge tubes serves sixteen fuel channels.

The position of the control rods in the moderator has been altered from our basic design. Instead of being pushed in and out from the sides, they operate in special vertical channels. These channels are directly beneath the charge tubes which pass through the roof, and this enables us to put a winding mechanism on the top of the tube and hang the control rod from it on a steel cable. The winding mechanism raises or lowers the control rods in the reactor core and so enables us to regulate the power level. In an emergency, the control rods fall into the core under their own weight and we do not therefore need a separate set of safety rods.

All these atomic power station reactors are producing not only heat but plutonium as well. This inevitable by-product is an artificial nuclear fuel of great value. Every industry tries to find a use for its by-products so as to reduce the cost of its main product, and in the same way we want to find the best means of using this plutonium which is formed in our reactors.

One way, of course, is to mix it with uranium which has lost some of its uranium 235 atoms—the fuel atoms which can be split—so as to restore it to its original standard as a fuel. If, however, you have a highly concentrated fuel, it is generally a bad thing economically to dilute it down: it is far more economical to design a machine which can take advantage of its special properties, and this is what we have done.

At Dounreay, in the north of Scotland, we have built a reactor which is designed to work with a very rich fuel like plutonium. The Calder Hall reactors use uranium as it is extracted from its ore—what we call natural uranium—and less than one percent of this can act as fuel. Because these fuel atoms—uranium 235—are so few we have to provide a large mass of graphite to slow down the neutrons so as to ensure that a sufficient number of the neutrons do manage to split the rare fuel atoms, and do not get captured instead by the atoms of uranium 238. If all or most of the atoms present are fuel atoms, we do not have to worry about the neutrons being captured, and so we need not provide any moderator to slow them down. This means we can get rid of over a thousand tons of graphite, and put our fuel close together. This makes the core of our reactor very much smaller. Instead of being something like the size of a house, it is now the size of a dustbin.

You will remember that the two and a half neutrons produced on the average each time an atom is split can go one of three ways. One from each fission must cause another atom to split to keep the chain reaction going; some are wasted because they are caught by materials in the reactor and get lost, and the rest are used to convert atoms of uranium 238 to plutonium.

In this plutonium-fueled reactor, there is much less material in which our neutrons can get lost, so there are more neutrons available to convert atoms of uranium 238 to plutonium. There is, of course, no uranium 238 in the core, so we surround the core with rods of this uranium and these capture the spare neutrons. Of the neutrons produced from each fission one keeps the chain reaction going by splitting another plutonium atom, on average about a quarter of a neutron is lost, leaving one and a quarter neutrons to escape from the core and convert uranium 238 to plutonium. So each time we split a plutonium atom in the core we get one and a quarter new plutonium atoms produced; that is, for every four atoms we burn, we get five new ones. In fact, we make new plutonium faster than we burn it. We call this breeding. Also because we do not slow the neutrons down, but use them when they are still traveling fast, it becomes a fast fission reactor. This combined with its ability to breed more new fuel than it burns makes it a fast fission breeder reactor.

Now how do we ensure that the spare neutrons do not go on hitting more atoms of plutonium instead of leaving the core and converting the uranium round the outside into fresh fuel? The answer lies in the precise arrangement of the plutonium in the core arrived at by a combination of intricate mathematical calculations and skilful design.

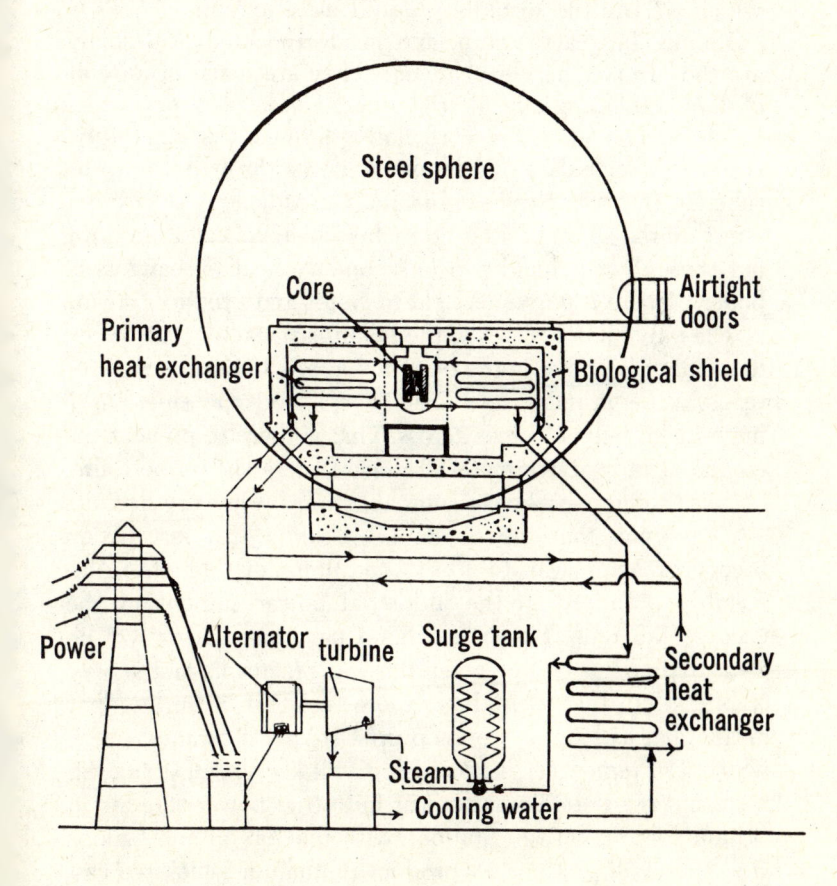

So in this reactor (Figure II-20) we have a core consisting of plutonium fuel elements surrounded by rods of uranium 238. Because the fuel is richer and closer together, the heat produced is much more concentrated than it is in the Calder Hall type of reactor. In fact, it is so concentrated that no gas could remove the heat fast enough, and we have to use a liquid metal. This liquid metal is a mixture of sodium and potassium. The mixture is pumped through the core of the reactor where it collects heat from the fuel. This heat we use to boil water and produce steam for our turbines just as in the other atomic power stations. Unfortunately, while the liquid metal is collecting the heat it becomes radioactive as well, so we cannot allow it to come outside the biological shield. Also, for various technical reasons, we do not want to take the water inside the biological shield. So in order to transfer the heat from the liquid metal to the water we have to provide a link between the two in the form of a completely separate circuit or loop of this sodium-potassium mixture, which we call the secondary liquid metal circuit.

Round the reactor core, and inside the biological shield, are the primary heat exchangers. They are made up of coils of double tubes, one inside the other. The liquid metal which has been heated in the core flows through the inner tube, while the secondary liquid metal flows through the outer tube. In this way the heat but not the radioactivity is transferred to the secondary liquid metal which we can then bring out through the shield to the secondary heat exchangers or boilers. There it passes its heat to water and produces steam.

The only other feature of the Dounreay reactor that I want to mention is probably the best known of all. Surrounding the reactor, the primary heat exchangers, and the shield is a huge metal ball, 135 feet across. That is there to prevent the escape of radioactive material in case there should be a fire, because sodium is very inflammable and burns very fiercely.

I told you that the object of building this fast fission breeder reactor was to burn the plutonium which is inevitably produced in the industrial power stations of the Calder Hall type. But from what I have just said it will be clear that once we have given this fast reactor its first supply of plutonium fuel it produces more fuel than it burns. That means that as long as we keep up a supply of uranium 238, which you remember is not fuel, it will keep itself going. So we can see a possible pattern of industrial power stations in Britain for the future. Stations with reactors of the Calder Hall type will produce by-product plutonium which we can use to start up an increasing number of fast reactor stations, which will then keep themselves going indefinitely. But there is another important advantage to be gained from building these reactors—this time an economic one. They enable us to burn up all our uranium, instead of only a small fraction, since the uranium 238—the atoms which will not split—can all eventually be converted to plutonium and used as fuel.

Indeed we can go further. There is another natural element —thorium—which can be converted into nuclear fuel. Its atoms will not split at all, but like atoms of uranium 238, thorium atoms can capture neutrons and change into another isotope of uranium—uranium 233 which is also an artificial nuclear fuel. So we could surround the core with thorium rods in place of the uranium 238 and still breed new fuel. This means that the fast breeder reactor will enable us to add all the earth's store of thorium to the stocks of uranium as a potential source of fuel for atomic power stations.

In saying this I am looking quite a way into the future. There are many difficult problems yet to be solved, though the reactor at Dounreay in the north of Scotland—which is really a big experiment—should give us a lot of the answers.

During this period of development, improvements to the Calder Hall type of reactor, which are continually being made, will provide a reliable and efficient source of heat for nuclear power stations, and of course that kind of reactor has the great merit that it runs on natural uranium.

The two types of reactor I have described so far are typical of the large range of possible systems, all of which work on the fission principle—the production of heat by splitting atoms. We can, however, also release heat by the opposite process—building atoms up instead of splitting them down —what we call the fusion process. The simplest of these fusion processes is the joining together of two nuclei of heavy hydrogen to form a nucleus of helium and a spare escaping neutron.

An atom of heavy hydrogen contains one proton and one neutron and when two of these atoms collide at very high speed their nuclei fuse together for a moment and form one

140 The Age of the Machine

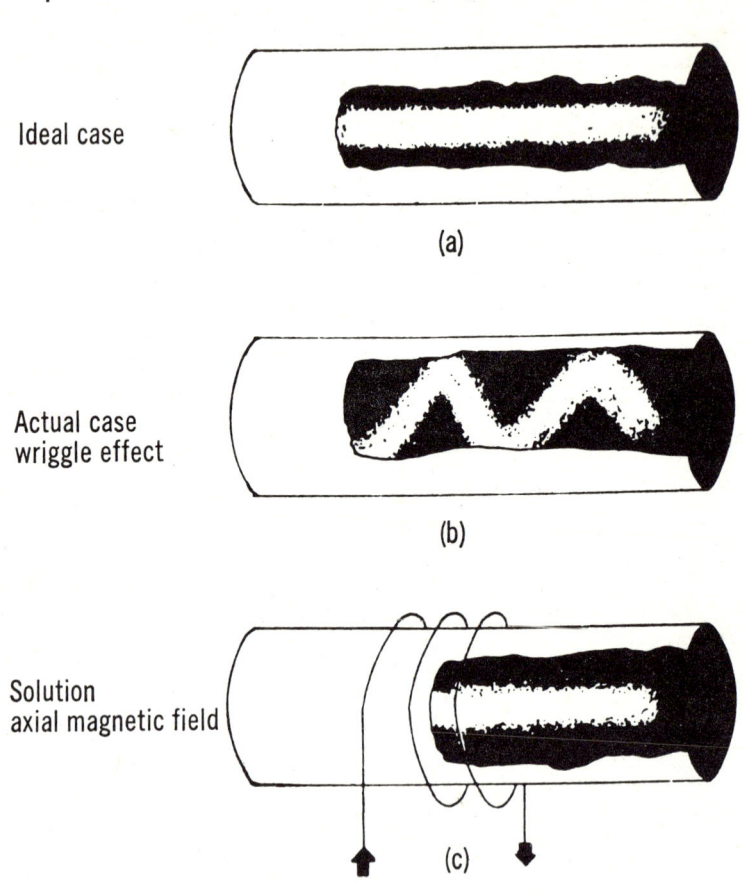

Ideal case

(a)

Actual case
wriggle effect

(b)

Solution
axial magnetic field

(c)

II-21.

atom of helium. This atom of helium contains the two protons from the heavy hydrogen atoms but only one of the neutrons. The other escapes.

The difficulty with this process is that the atoms have to be traveling at enormous speeds before their nuclei will fuse together. One way of speeding them up is to raise their temperature, but to get them going fast enough to fuse needs a temperature of several million degrees—far hotter than the surface of the sun. Clearly, we cannot reach this sort of temperature, or anything approaching it, by heating the atoms over a fire or in a furnace since all known materials would melt and vaporize at temperatures far lower than this, and so we have to find some other way.

The most promising way, it seems, is to pass a powerful electric current through heavy hydrogen gas in a tube. By this means the gas can be made extremely hot but there is the further task of maintaining this temperature long enough for fusion reactions to take place. Some very successful work has already been done along these lines in both the United States and Britain, and in January 1958, Harwell announced for the first time that they were able to heat up heavy hydrogen gas to a temperature of five million degrees and hold it for a few thousandths of a second.

How was this done? British scientists and engineers at Harwell built a large apparatus called Zeta. It looks like a big transformer the size of a double decker bus. The vital part of Zeta is a large container shaped like the inner tube of a motorcar tire; this tube is about ten feet across and has thick aluminium walls. Heavy hydrogen gas is put into the tube and a jolt of electricity is passed through it from a transformer. The object of this electric current is to heat the gas to a temperature at which the atoms of heavy hydrogen will fuse together.

Fortunately as soon as the electric current begins to flow the gas is automatically squeezed into a narrow column in the middle of the tube away from the walls [Figure II-21(a)]. This happens because all electric currents have magnetic forces acting round them, and in a gas these magnetic forces pinch the gas towards the center of the tube.

Up to this point nature works in favor of the scientist. But as soon as the gas is squeezed like this, we face a major problem. The gas carrying the electric current immediately begins to wriggle and jump about inside the tube [Figure II-21(b)]. Unless we can do something about this wriggle it will touch the walls and lose its heat and perhaps melt the wall. It is essential to keep the gas in the center of the tube for as long as the electric current lasts.

In Zeta the gas is kept centrally in the tube and well away

from the walls by imposing additional magnetic forces along the path of the current [Figure II-21(c)]. These magnetic forces are generated by winding coils of wire round the outside of the aluminum tube and then passing electricity through the coils. When this is done the wriggle disappears.

The main achievement with Zeta has been to hold the gas in the center of the tube long enough, and at temperatures high enough, to observe fusion taking place. Now how can we observe fusion?

Well, remember the fusion of two heavy hydrogen atoms gives an atom of helium containing three atomic particles in its nucleus, and in addition a spare neutron. These spare neutrons escape from the tube altogether and it is by detecting them that we know that fusion must be taking place in Zeta.

The next step facing research scientists is to raise the temperature of the gas even higher by putting still more electrical energy into Zeta and its successors. At higher temperatures held for longer times the amount of energy produced from fusion of the gas will increase until it exceeds the amount of energy needed to create the tremendous heat. We shall then start showing a profit, and ultimately be able to convert this energy gain into electrical power. But scientists estimate that in order to show the profit we are after, we shall need to heat the gas to temperatures exceeding a hundred million degrees.

So the main problems facing scientists are heating the gas to a temperature above a hundred million degrees, holding it at that temperature for much longer times, and finding ways of converting the energy gain into electricity. No one can say just how long it is going to take them to crack these enormous nuts—it may well be many years. And even when these major problems have been solved, many engineering difficulties will have to be overcome before we see the fusion equivalent of Calder Hall. (Figure II-22).

What then is the advantage of the fusion reaction which justifies all this elaborate research? The answer lies in the fuel it burns. We get it from ordinary water. Water is made up of oxygen and hydrogen and I said earlier that most hydrogen atoms have one proton but no neutron, but that one atom in every six thousand is different. It has both a proton and a neutron and is called an atom of heavy hydrogen or deuterium. This in fact is our fuel. It is obtained by a separation process costing only a few shillings a gram to produce. The energy stored in this thimbleful of deuterium is equivalent to the energy stored in several tons of coal. So the fuel is not only cheap—its supply is inexhaustible.

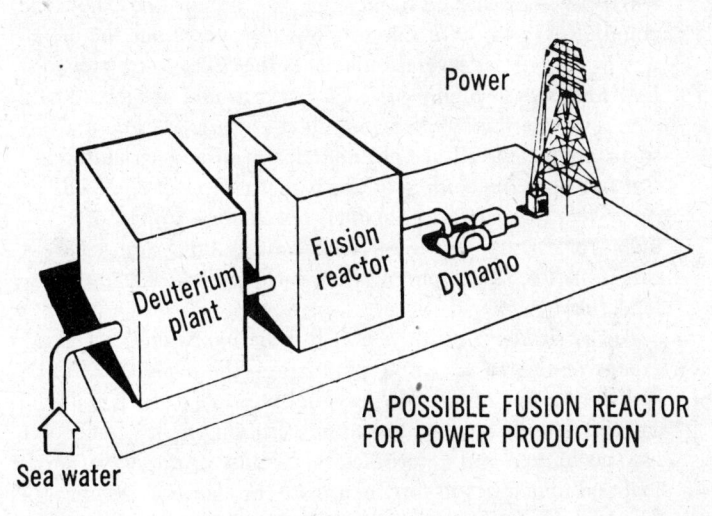

A POSSIBLE FUSION REACTOR FOR POWER PRODUCTION

II-22.

B. Materials and Methods

Archaeologists divide man's history into three main divisions —the Stone Age, the Bronze Age, and the Iron Age—each with many subdivisions and frequently overlapping. It is possible that the first use of iron began during the Stone Age when it was discovered in meteors. We do not know how the complicated process of smelting was discovered but the metal was in limited use several millenia before Christ. In wrought-iron form it was highly prized for swords and daggers. Gradually, as tempering by quenching in cold water became understood, its use spread in both craftsmanship and agriculture. But not until the coming of the blast furnaces of the Middle Ages were temperatures obtained hot enough to melt the metal for casting. In the fifteenth century, iron cannons were cast in molds, thus sounding the death knell of chivalry and feudal castles.

In the sixteenth century, iron became increasingly scarce and expensive, due to increasing population, the demands of war, and the scarcity of charcoal for smelting. In Queen Elizabeth's day this scarcity resulted in the substitution of coal, but such use was limited and unsatisfactory, because of impurities in the coal and difficulties in forcing a blast through the mixture. Abraham Darby of Birmingham was the first to use coke in blast furnaces, and founded a family of great ironmasters. Coal mining boomed and the problem of freeing the mines from water led to the steam pumps of Savery, Newcomen, and Watt. The connection between smelting and steam power became closer when John Wilkinson used the new invention to create a hotter and more continuous blast independent of water. He ordered his machine from the newly founded firm of Boulton and Watt in 1776. Another improvement had been made by Benjamin Huntsman, a Doncaster watchmaker, who produced a steel free of slag and other impurities in the intense heat of a crucible furnace. He moved to Sheffield in 1744, where a competitor dressed as a beggar is said to have *learned his secret. His process, at first rejected by the city's cutlery makers, was in common use for this purpose during the following decades. Nevertheless, the manufacture of cast steel continued to be a costly process and wrought iron, although inferior in performance, remained the material in most common use. The innovations of three engineers of genius, whose contributions are described by J. D. Bernal, introduced the fourth great age of man—the Age of Steel—in which we now live. Dr. Bernal, a noted crystallographer, is Professor of Physics at the University of London, and author of several books including* The Social Function of Science *and* Science and Industry in the 19th Century, *from which this selection is taken.*

THE AGE OF STEEL

J. D. BERNAL

IN THE LAST THIRD of the nineteenth century, economy and politics were largely influenced by the availability, cheap and in quantity, of what was virtually a new material—cast steel, which could be rolled or forged. Steel rapidly replaced iron in most of its important large-scale uses—for rails, for ships, for bridges and for buildings. Its superior strength and lightness not only made it cheaper for these purposes but also enabled much more daring constructions to be attempted. The adoption of steel was a major factor in opening the whole world to a new level of trade and exploitation in a matter of decades. It made colonial dependencies and undeveloped areas much more profitable, helped capital expansion, and was the material basis of the new imperialism of the end of the century. Economic and technical factors combined to make the steel age the prologue of a new period of wars and revolutions.

The transition from iron to steel appears on close examination to be due to the work of far fewer men than any other major transformation in techniques. Three great names stand out clearly, Bessemer of the converter, Siemens of the regenerative open hearth, and Gilchrist Thomas of the basic lining. The three changes they introduced, though extremely simple in underlying principle, marked each in its own way an abrupt and decisive change in technique. Yet they were more than technical changes, as witness the fact that despite the existence of a growing and enterprising iron and steel industry, not one of them came from anyone in the trade but instead from rank outsiders, whose ignorance of what could not be done was their chief asset. The technical break involved in large-scale steel production was one too great to be bridged by slow improvements of existing techniques; it needed fresh and imaginative minds that did not know what could not be done.

Bessemer was an inventor and manufacturer of bronze powder; Siemens was an engineer whose chief interest had been in electricity and the steam engine; Thomas was a police court clerk with a classical education and a hobby of chemistry. What made their achievements possible was that they all, in very different degrees, made use of science. But this, in itself, is no explanation. There were, in their time, hundreds of engineers and chemists who knew far more science and even more of the science of metallurgy than they did. What they had in common, and what made their success, was their appreciation of the existence of technical and economic problems to which it was worth applying science. They were determined to solve them by a radical examination of the ends to be achieved and not merely, as most metallurgists had done before them, by a mere modification of existing practice.

The approach of each of the three was very different, characteristic not only of the man but of his times. Bessemer, born in 1822, was an unusual combination of prolific inventor and self-confident enterprising capitalist of the mid-century. Sir W. Siemens, a trained engineer, was one of a band of brothers determined to apply science in one form or another to make their fortunes, characteristic of the new

German capitalism of the seventies. Gilchrist Thomas, solitary and poor, was the only one of the three who appears to have been conscious of exactly what he was trying to do. He alone analysed the whole problem and found a theoretical solution which was successful in practice almost from the start. What is perhaps even more remarkable is that, unlike such inventors in the past, he had secured, in advance, a complete cover of patents, which even the most powerful interests were unable to break. In his work one can see the prototype of the industrial research establishment of the next century.

By 1854 the manufacture of iron, cast and wrought, was expanding with ever increasing speed, but steel was still a luxury product, selling at about £50 a ton, of very uncertain quality and available in small pieces, as against £3–£4 for pig iron, and £8–£9 for wrought-iron rails. It might have seemed that this disparity in price between products containing approximately the same materials would have led to an intense search for new methods of manufacture, but rightly or wrongly no ironmaster thought it was worth his while. The steelmakers on their side were as satisfied with a small turnover of a valuable commodity.

Henry Bessemer entered the field almost by accident, but was well equipped for the task. The son of an engineer and craftsman, he had little formal education but great practical experience, particularly of type casting and the reproduction of metals and works of art on which he supported himself in his early years. He had a genuine passion for mechanical work and had picked up a good deal of science, particularly chemistry. At the age of thirty he had had the idea of replacing the age-old hand process of making powder for gold paint out of leaf brass by one using machinery alone. What led him to do so is best told in his own words:

> My eldest sister was a very clever painter in water colors, and in her early life, in the little village of Charlton, she had ample opportunities of indulging her taste for flower painting . . . and had, with much ingenuity, made

a most tastefully decorated portfolio for their reception. She wished to have the words—

STUDIES OF FLOWERS
FROM NATURE,
by
Miss Bessemer,

written in bold printing letters within a wreath of acorns and oak leaves which she had painted on the outside of the portfolio; as I was somewhat of an expert in writing ornamental characters, she asked me to do this for her, and handed me the portfolio to take home with me for that purpose.

How trivial and how very unimportant this incident must appear to my readers. It was, nevertheless, fraught with the most momentous consequences to me; in fact, it changed the whole current of my life, and rendered possible that still greater change which the iron and steel industry of the world has undergone, and with it the fortunes of hundreds of persons who have been directly, or indirectly, affected by it.

The portfolio was so prettily finished that I did not like to write the desired inscription in common ink; and as I had seen, on one occasion, some gold powder used by japanners, it struck me that this would be a very appropriate material for the lettering I had undertaken.

How distinctly I remember going to the shop of a Mr. Clark, a colourman in St. John Street, Clerkenwell, to purchase this "gold powder." He showed me samples of two colours, which I approved. The material was not called "gold" but "bronze" powder, and I ordered an ounce of each shade of colour, for which I was to call on the following day. I did so, and was greatly astonished to find that I had to pay seven shillings per ounce for it.

On my way home, I could not help asking myself, over and over again, "How can this simple metallic powder cost so much money?" for there cannot be gold enough in it, even at that price, to give it this beautiful rich colour. It is, probably, only a better sort of brass; and for brass in almost any conceivable form, seven shillings per ounce is a marvellous price.

. . . Here was powdered brass selling retail at £5 12s.

per pound, while the raw material from which it was made cost probably no more than sixpence. "It must, surely," I thought, "be made slowly and laboriously, by some old-fashioned hand process; and if so, it offers a splendid opportunity for any mechanic who can devise a machine capable of producing it simply by power."

I adopted this view of the case with that eagerness for novel inventions which my surroundings had so strongly favoured, and I plunged headlong into this new and deeply interesting subject.[1]

Realizing that everything depended on secrecy he devised an almost automatic plant which he set up in a factory without windows, operated only by his wife's three brothers. Thus he retained his secret for forty years. This achievement was in many ways more difficult than that of making steel and points to the factories of the age of mass production. As a result he made a small fortune, which he found essential for his large-scale experiments on making steel. He had, in fact, financed his own industrial research establishment.

Always combative, particularly when roused by official complacency and stupidity, he determined to attack the problem of the large-scale production of cast steel just because the War Office had maintained that it could not be done. He had devised an aerodynamically spun shell adapted to smooth-bore cannon which they said no cast iron gun could fire. He inquired why a steel gun could not be used and was told condescendingly that no such gun could be made, because uniform steel could not be had in quantity. He thereupon decided to produce it. His only qualification for the enterprise was his successful development, a few years before, of a method of continuous casting of glass sheet which is the basis of all present methods. This gave

1. It was from old books, probably based on the *Treatise upon Divers Arts* by Theophilus the monk in the eleventh century, or a derivative source that Bessemer found the principles of the process which was still carried out on a traditional basis in Germany. See *Sir Henry Bessemer, F.R.S., An Autobiography*, London, 1905.

him a good knowledge of the making and management of furnaces. In his own words:

> My knowledge of iron metallurgy was at that time very limited, and consisted only of such facts as an engineer must necessarily observe in the foundry or smith's shop; but this was in one sense an advantage to me, for I had nothing to unlearn. My mind was open and free to receive any new impressions, without having to struggle against the bias which a lifelong practice of routine operations cannot fail more or less to create.

He started with the idea of using the first scientific method of steelmaking, devised by Réaumur in 1720, of fusing together wrought iron, almost carbon free, and cast iron, saturated with carbon. This he proposed to do by using a reverberating furnace, but had little success. Noticing that much of the gas remained unburnt, he introduced air pipes into the furnace itself. This increased the flame temperature but also, he noticed, burnt the carbon out of the iron. Seizing on this observation he was led, step by step, to blowing air through the molten metal in fixed and then in movable furnaces, and ultimately to his famous *converter*. This development, which took barely two years, was an admirable example of the combination of observation and full-scale experimentation with scientific explanation of each step. In 1856 he succeeded in turning out the first ton of cast steel from his converter and announced his success, characteristically enough for the times and the man, at the Cheltenham meeting of the British Association.

Bessemer's process was enthusiastically taken up by the ironmasters, but almost as quickly dropped because of the apparently fatal flaw of failing to remove the phosphorus which most English and European ores contained. He himself, however, persisted and set up a works in Sheffield, and in the teeth of the steelmakers made steel with phosphorus-free ores from Sweden, Spain, or Cumberland which was better than most and cheaper than any that they made. The rest of Bessemer's business life was spent in fighting a long, but ultimately successful, battle against the prejudices and interests opposed to the use of the new steel. This, though it involved little science, was an essential part in the changeover from iron to steel. In the hands of a less combative and businesslike man, the innovation might have been overborne and the transformation postponed for decades.

This surmise requires the usual qualification that inventions rarely come singly. About the same time a number of men were working towards the idea of violent oxidation of molten cast iron. Nasmyth had already in 1854 blown steam through molten iron to keep it stirred; he was within an ace of discovering the converter but he generously admitted Bessemer's claim to the full credit because he understood what he was doing. William Kelly, a Pittsburgh man, came closer still and made what would be now known as a side-blown converter in his Suwanee Furnace on the Cumberland river in which he made plates for the Ohio and Mississippi steamboats as far back as 1851. Lacking the knowledge, the facilities, and the money that Bessemer could command, and caught by the instabilities of American finance, he was unable to develop his method and his patent was invalidated.

The practical success of the converter was assured once it became possible to control accurately the low but vital carbon content of the steel it poured. One way to achieve this, but a difficult and unreliable one, was to stop the blast at the right moment. A far better way was to burn all the carbon out and then to restore just the right amount at the end of the process. Spiegeleisen, an iron manganese carbide, proved to be an ideal way of doing this, as the manganese had an additional deoxidizing effect and cured the defect of early Bessemer steel of being red-short. Its preparation and properties had been investigated by a skilled metallurgist, Robert Mushet, who proposed its use in steel making. He was the son of David Mushet, the discoverer of the black-band ironstone that was to make the fortune of the Scottish iron industry. Mushet's patent was badly drawn up and Bessemer was able to evade it, but he seems to have had some conscience in the subject for on being appealed to by Mushet's daughter he paid the now impoverished steelmaker a pension of £300 a year until his death in 1891. There is no question, however, that it is to Bessemer and not to Kelly or Mushet that the real credit for making steel on a large scale is due. Yet, in spite of his early success, many obstacles still remained to be overcome before the full day of cheap steel.

To many ironmasters the converter seemed an unnecessarily costly and complex machine and they welcomed a simpler way of making steel in large quantities. This was provided in 1867 by the Siemens-Martin process. It is difficult to place Sir William Siemens in the history of science and technology because, though by himself he would be a typical scientific inventor of the nineteenth century, he cannot be considered apart from his family of six distinguished and enterprising brothers, particularly the eldest brother, Werner von Siemens, who saw to his education. It was the latter who directed what was virtually an international syndicate of inventors and industrialists centered in Germany but operating in Britain, France, Russia, and indeed all over the world. The Siemens family contributed perhaps more than any other group of persons to the creation of modern German industry and gave it, from the start, the academic scientific basis that neither British nor American industry acquired till very much later.

Nevertheless, the industrial level in Britain in the sixties and seventies was still higher than in Germany, and Sir William Siemens—the English Siemens—found there a wider market for the exploitation of the family's inventions. Though these were multiple and covered an enormous range, the most successful were the production of instruments and cable for the growing telegraph industry, the development of the dynamo and electric power production, and industrial furnaces. Unlike Bessemer, who attempted to solve a number of specific problems one after the other, using any means that occurred to him, what we now would call a series of converging researches, the Siemens were consciously applying certain general engineering principles and feeling their way as to where they could be used most profitably, typical divergent research.

The principle that concerns us here is that of regeneration. The idea is one of common sense backed by the thermodynamic principle that heat should not be allowed to escape from any industrial process but should be used for warming up incoming products. Applied to metal furnaces it meant an end of the flaming chimneys that for nearly a century had lit up the sky in Sheffield and the Black Country. Instead the hot gases heated a mass of bricks white-hot and escaped when relatively cool, while fresh air was later heated up by forcing it over the hot bricks. Originally intended as a means of economizing fuel, regenerative furnaces proved to have the unhoped-for power of raising temperatures to levels never before reached. Using gas fuel to avoid contamination they made the original Réaumur process of a mix of scrap iron and pig iron possible on a large scale and on an open hearth. Just that extra temperature was needed to convert the tedious puddling process, in which the purer, and consequently less fusible, iron was drawn out in a pasty condition, into one in which the whole charge could become a bath of melted steel which could be poured out by the ton.

There were, naturally, many practical difficulties, particularly in finding refractory furnace linings, but these were largely overcome by the French ironmaster, Etienne Martin. The final result by 1867 was a process as cheap and good as Bessemer's, but slower and far less mechanically and chemically revolutionary, thus appealing more to the traditionally minded steelmakers. Siemens himself was not notably successful in his use of it. His own Landore works, though it turned out excellent steel, suffered seriously in the depression of the seventies and Sir William lost there much of what he had gained in his other enterprises.

The regenerative open-hearth furnace did not depend as much as Bessemer's on phosphorus-free ore, but it still could not make steel out of the highly phosphatic ores of Cleveland and the by then German Lorraine, which could only be used for iron. By the middle of the seventies this disparity was a glaring one and the problem of making steel from phosphatic ores was an open challenge. Yet despite all the new talent called into action in the steel industries of Britain, France, and Germany the solution did not come from any ironmaster or metallurgist, but as the result of the deliberate and carefully directed researches of a very part-time scientist, Sidney Gilchrist Thomas.

He came from a mixed Welsh and Scottish family with serious and intellectual tastes, particularly marked in his mother. His father, a civil servant, was never very well off and his early death left the family in very reduced circumstances. It was for that reason that Sidney, at seventeen, with a promising university career before him, felt he must give it up and obtain an extremely wearing job as a junior police court clerk in the London docks, the conditions of which probably conduced to his early death from tuberculosis. He continued his studies at Birkbeck College in the evenings, concentrating on chemistry. There he heard the lecturer, Dr. Chaloner, say that anyone who could solve the problem of making steel from phosphoric ores would make his fortune. Thomas needed that fortune badly, not, as might be thought, for the support of his mother and sisters, but because the horrors which he had to witness every day at his East End court filled him with idealistic schemes for social reform.

He, therefore, set himself deliberately to the problem, not as Bessemer had by observations based on large-scale trials, but by scientific analysis of the factors involved. He read voraciously in the technical literature and himself became a contributor to the journal *Iron*, using his scant leisure to visit ironworks at home and abroad. The solution of the problem which had baffled all professional metallurgists took him from 1871 to 1875, using a cellar as a laboratory and working in what time was left between his duties and his courses.

Essentially the solution was that of absorbing the oxidized phosphorus or phosphoric acid in a basic lining, made from magnesian limestone. The necessary practical tests were carried out in the course of 1877 by his cousin, Percy Gilchrist, who was employed as a chemist in the Blaenavon ironworks, working at first with a miniature converter holding eight pounds. Later with the assistance of the manager, Mr. Martin, they worked up to 11 cwt. By March of the next year Thomas was confident enough to take out patents and announce his discovery to the Iron and Steel Institute, where no notice whatever was taken of it. Again in September, at the Paris Exhibition of 1878, his paper on the new method was not found sufficiently interesting to be read, but there Thomas met Mr. Richards, the manager of a Cleveland ironworks, and impressed him enough to persuade him to make some full-scale tests. Finally on April 4th, 1879, the first few tons of steel were made at Middlesbrough.

Immediately the whole steel world was at his door clamoring for licenses. Krupps tried to evade the patents, but Thomas, who had not been a police court clerk for nothing, brought them to law and forced them to pay up. The process, unlike Bessemer's, was an almost unqualified success from the start. It made the steel industry of Middlesbrough, of the Ruhr, and of the southern United States. The steel age had begun in earnest, but the man who had made that possible and who could have gone on to further triumphs had only a few years to live, dying in Paris in 1885 and leaving the bulk of a now considerable fortune to charity.

In considering these three contributions to the nineteenth-century revolution in metallurgy, the first striking point is their independence of any organized scientific movement. Of the three inventors only Siemens had a university education, and none of them received any material assistance or more than a little advice from academic, scientific or governmental sources. This is the more surprising as by the midcentury the official scientific movement was in full swing, thanks in part to the German influence that came in with Prince Albert. In fact, the beginning of organized and subsidized industrial research had to wait another ten years, though its value had been amply proved by Bessemer's work.

In those days it was still thought sufficient to provide institutes and bodies like the British Association for promoting and discussing the reports of research carried out by private means. The inventor was expected to recoup himself by exploiting his invention under cover of the patent laws, if he was lucky. We see, in fact, Bessemer and Siemens using the results of one invention to finance the next, while Thomas was only able to work his way up in the first place by pinching and starving himself. Even the military departments, which were to benefit so largely by the results of the revolution in steelmaking, did not encourage it and indeed were so attached to older materials and methods that they subjected the innovators, particularly Bessemer, to a series of painful snubs. These conditions were exaggerated in an industry like iron and steel, which had grown up in the hard way by expanding the operations of smiths, with the occasional injection of critical new ideas by outsiders, such as Roebuck and Cort in the eighteenth century, both of whom ruined themselves for their pains.

The result of this state of affairs was that new methods penetrated very slowly and there was an altogether unnecessary time lag between scientific knowledge and its application. None of the scientific ideas used by Bessemer, Siemens and Thomas were more recent than 1790 and the steel age might well have begun fifty years before it did. It would, however, have grown more slowly, owing to the far more limited technical and financial structure of capitalism early in the century. Equally well, but for the enterprise and good fortune of the inventors, particularly Bessemer, it might have been delayed, but not for so long, because it is clear from the methods of Siemens, and still more of Thomas, that science was soon to reach a stage when it could be used consciously and deliberately to solve industrial problems.

The growth of engineering has been fostered not only by such inventions as those of Watt and Bessemer, but by innumerable less revolutionary devices and materials which have become part of the standard equipment of the profession. In the history of construction, one of the most interesting and useful of such materials was developed during the 1840's by a German immigrant to the United States named John A. Roebling, who was to become one of the country's greatest bridge builders. Roebling was born in Prussia in 1806, studied at the Polytechnic School in Berlin and after emigrating in 1831, set up shop as a civil engineer in Western Pennsylvania. Here he developed his new process for the manufacture of wire rope. His original paper describing the invention appeared in the "American Railroad Journal and Mechanics' Magazine" in 1843. The first important use of the material was in the Grand Trunk Bridge at Niagara Falls, which Roebling undertook to construct after another engineer had resigned. The bridge, for railroad use, was an unqualified success, and inaugurated an active era of suspension bridge building in the United States. Roebling's most spectacular achievement, shared with his son Washington, was the construction of the Brooklyn Bridge. Part of the story of that monumental undertaking is reprinted on page 255.

MANUFACTURE OF WIRE ROPES

JOHN A. ROEBLING

THE ART of manufacturing ropes of wire is comparatively new. Numerous attempts have been made in Europe and here, and most of them have proved failures. A collection of parallel wires, bound together by wrappings in the manner of a suspension cable, is no rope and not fit for running, it can only be used for a stationary purpose. The first rigging made in England was of this description. The difficulty in the formation of wire rope arises from the unyielding nature of the material; iron fibres cannot be twisted like hemp, cotton, or woolen; their texture would be injured by the attempt. To remove this obstacle, some manufacturers have resorted to annealing, and thereby destroyed the most valuable properties of iron wire, viz. its great strength and elasticity.

My first attempts in the manufacture of wire rope were made four years ago, my intimacy with the construction of wire cable bridges induced me to investigate this matter. The principles of my process differ from those of the English manufacturers, they are original and secured by patent. The novelty of my proceeding chiefly consists in the spiral laying of the wires around a common axis *without twisting the fibres,* and secondly in subjecting the individual wires while thus laying to a uniform and forcible tension under all circumstances. By this method, the greatest strength is obtained by the least amount of material, and at the same time a high degree of pliability. Each individual wire occupies exactly the same position throughout the length of a strand; another result of the precision and force applied in laying is the close contact of the wires, which renders the admission of air and moisture impossible.

Three years ago I offered to the board of canal commissioners, which was then in power in Pennsylvania, to put a wire rope of my manufacture on one of the planes of the Allegheny Portage railroad, at my own risk and expense, the value of the rope to be paid in proportion as it rendered services equivalent to a hemp rope. This liberal offer however was rejected and not considered until the present board came into office. Last year I put three ropes, measuring in the aggregate 3400 feet, 4½ inches circumference, in operation on plane No. 3. Owing to the want of adhesion, I had at the start to contend with some difficulties. By means of a double groove on the receiving sheave and a guide sheave placed back of it, which crosses the rope and leads it from one groove to the other, which improvements were added to the machinery last winter, I succeeded in doubling the adhesion. When in unfavorable weather there is delay and slipping on the other planes the wire rope can at all times pull as heavy a load without a balance, as the engine is capable of hauling. The planes of the Portage railroad require hemp ropes of 8½ inches circumference, made of the best Russian or Italian hemp, and which cannot be trusted longer with safety than one season. They are frequently from reasons of economy continued 1¼ seasons; much however, depends upon the weather and business. The unfavorable circumstances under which the wire rope had to work last year affected it some; the wear of the whole of this season, however, is not perceptible, and its present condition promises a long duration. I am now manufacturing another wire rope of 5100 feet long in four pieces for plane No. 10.

The first rope of my make, 600 feet long, 3¼ inches circumference has now been in successful operation two seasons, hauling section boats from the basin to the railroad at Johnstown. Two more were put to work last spring at the new slips at Hallidaysburg and Columbia. From my present experience, I may safely assert that wire rope deserves the preference over hemp rope in all situations much exposed, and where great strength and durability is required.

By my process of manufacture, the same pliability is imparted to the rope which is proper to the wire itself. Paradox as this may appear, it is nevertheless a fact and is easily explained. By pliability is here understood the extent of flexure to which the rope or wire may be subjected, without producing a permanent bend; when released the rope must resume its former and straight position. To bend a rope requires force, and this force is in proportion to its areal section, *caeteris paribus*.

Well manufactured iron rope is more pliable than hemp rope of the same strength. I am manufacturing tillers of fine wire, capable of bearing 3000 pounds and which ply around cylinders as small as four inches in diameter, and in which the wires are so compactly laid, that not the slightest shifting in their spiral position is to be observed. A number of my tillers are in use on the Ohio and Mississippi. Such rope would be pliable enough for running rigging and be of long duration.

I will here add a few remarks on the introduction of standing wire rigging in place of hemp rigging. This subject has, for some years past, engaged the attention of the navy department of England and France, and the success which has attended the use of wire in place of hemp for shrouds and stays in the naval and commercial service of Great Britain, would, it appears, seem to warrant an attempt to test its utility on our national vessels.

Allow me to cite here a few remarks from the notes of Capt. Basil Hall, on a tour through Switzerland, and while examining the wire suspension bridge at Fryburg. He says,

> attempts are now making and will ere long succeed, to introduce wire rigging, which is stronger and better than chain, because less dependent on the accidental quality and careless manufacture of a single part. How strange it is, that the plan of making wire bridges, so successfully adopted in France, and elsewhere, should not have found favor enough in England to be fairly tried on a large scale. Fryburg bridge 301 feet wider than Menai, at least equally strong, has cost only one-fifth of the money. I do not think wire will answer for running rope; but for standing rigging it may, I conceive, be most usefully substituted for hemp, for with equal strength experience shows it to be lighter.

The cables of suspension bridges are stationary and will, when protected against oxidation, last an indefinite period. Standing rigging (when compared to running rope) is nearly stationary, and there is little wear but what arises from the direct strain, which if supported by sufficient strength, will have no deteriorating effect. In comparing the two materials, wire and hemp, for rigging, the state of preservation and time of use should be considered. For instance, a hemp stay of a certain size, made of the best Italian hemp, will when new, possess two-thirds of the strength of a wire stay of the same weight per foot; but let the two stays have been exposed and served five years, then the strength of the hemp stay will be gone, while the wire stay will not show any perceptible reduction. In this case, of course, a common wear and tear is supposed.

The most prominent features of wire rigging, as compared to hemp rigging, are its great durability, less weight and size, less surface exposed to wind, less danger in time of action of being destroyed by shot. Another good quality of the wire rope is its great elasticity which is quite sufficient to counteract the effect of a sudden jerk, while a vessel is rolling heavily at sea. The elasticity of hemp rigging is only to be relied on to a very small extent; it will give and stretch a great deal but not return.

A common objection of those not familiar with the nature of wire rope is its supposed rapid destruction by oxidation, but no apprehension is less founded than this. Running wire rope while in use either in or out of water, in mines or any other situation, will not even require the protection of oil, varnish or tar; while at work it will rust no more than a rail or a chain in use, but when idle, oxidation will affect it in proportion to the surface exposed. As, however, the process of laying is carried on with mathematical precision, by which the wires are brought into the closest contact, the assemblage of wires in form of a strand present a solid rod, which will be no more subjected to rusting than the link of a chain of the same size. The individual wires as well as the strands and ropes are coated with an excellent varnish during the manufacture. Wire rigging will require no other protection but oiling or tarring once or twice a season. Where elegance is an object, black or green paint may be used. Rigging made of zinked wires and not painted, would present a most elegant appearance and be exempt from all rusting.

Wire rope can be spliced in the same manner as hemp rope. The attachment of wire shrouds to the sides of the vessel and to the masthead and their connection with the rattlines (which should also be of wire) can be effected by the old method; the use of wire however, will suggest some modifications better adapted to the material.

Some wire rigging has been manufactured in England which simply consists of a collection of parallel wires bound together and served over with hemp. These mixtures, as experience has proved in the case of tiller ropes, are objectionable, the wire will rust inside of the hemp in spite of all protection by varnish; besides, the cover of hemp, which adds nothing to the strength, is only an additional expense.

Iron is now gradually superseding wood in the construction of vessels; a complete revolution in ship building has already commenced in England. Although very expensive at first, iron ships will prove the cheapest in the end. Are there any well founded objections to wire rigging, which assumes the same relation to hemp rigging as wooden ships to iron ones? There are none. Why then not test this matter by encouraging those who are capable of bringing it to perfection? A number of iron vessels are now building for the naval and revenue service, which seem to offer appropriate occasions for the test of this matter.

The earliest practice of mining preceded the dawn of written history. Such metals as gold and copper, which occur in the free state, were mined before the science of metallurgy was born. As civilization advanced and demand increased, mining with pick and shovel, hammers and wedges, increased. Shattering of rock by heating and sudden cooling with cold water became accepted procedure. The first record of extensive mining by means of shafts beneath the surface seems to be that of the Greeks at the silver mines of Laurion. The Romans sought out metals with avidity, not only in Italy but in every other corner of the Empire. Although they permitted conquered territories a measure of self-government, they took strict control of mineral resources. Slave labor was regularly employed and the appalling working conditions can be imagined. Shafts and tunnels sank deeper, galleries were supported by timbers, the perennial problem of water in lower levels was met with bucket wheels and screw pumps. In general, no revolutionary methods were employed. Production judged by later standards was extremely small, but the gold, silver, copper, and quicksilver of Spain, the tin and lead of Britain, and iron from as far away as India played their part in shoring up the power of the Empire and adorning its inhabitants.

The greatest early treatise on mining, "De Re Metallica" by Georg Bauer, known as Agricola, was printed in 1556. It described machinery, techniques, and underground construction with such lucid text and illustrations that it remained a standard reference work for hundreds of years. Many of the methods advocated are still in use. Incidentally, a translation from Latin into English was published by former President Herbert Hoover and his wife in 1912.

Over the centuries, the muscles of men were gradually supplemented by horsepower and rudimentary pumps and stone crushers, and as we have seen a new era in both mining and metallurgy began with the steam engine. Profound changes in technique resulted from new discoveries in the pure sciences—particularly chemistry. Large-scale operations became the rule. In the past one hundred and fifty years the profession has seen changes as great as in all the period preceding them. The most important techniques now in use are described in this article by C. C. Furnas, a civil engineer who is now Chancellor of the University of Buffalo. It must be remembered, however, that mining is essentially a provincial enterprise and that techniques vary widely according to the nature of the mineral and the terrain in which it is found. The emphasis today is on automation devices, such as larger loading machines, higher speed rock drills, and greater use of belt conveyors. A belt system, now in full operation, carries iron ore from Cleveland to Pittsburgh and coal from Pittsburgh to Cleveland.

The depletion of natural resources, particularly iron, caused by two great wars has alerted the great powers to the importance of obtaining new sources of supply. More extensive exploration and more sophisticated location techniques have been adopted. New processes by which low-grade ores such as taconite can be profitably worked have been invented. Substitution of man-made plastics is another solution which is being investigated. Although we can do no more than guess when the event will take place, we may at some time in the future find ourselves in dire want of essential minerals. It becomes important for the survival of any nation that it make discovery and conservation of national resources a matter of urgent public policy.

DISCOVERING, RECOVERING, AND CONCENTRATING THE MINERALS

C. C. FURNAS

Discovery of Ores

OLD CACTUS PETE, the prospector, with long beard, slouch hat, a great deal of hope, and a faithful burro loaded to the gunnels with grub and pick and shovel, is almost a trademark for western America. Civilization owes much to men of the Cactus Pete type. Had there not been many optimists in the past few thousand years, who were possessed with the overpowering urge to search for hidden wealth, there would have been no civilization. Minerals are useless if no one knows where they are. Some of the old recipes for rabbit soup started out with the instruction: "First you catch your hare." The first part of a recipe for a civilization built on minerals is: "First you find your minerals."

The old-time prospector merely looked carefully to find what he could on or near the surface. He panned the sands of stream beds to see if there was a bit of the color of gold. He looked over the earthy debris from a woodchuck hole to see if the ignorant creature had discarded some nuggets. He broke off pieces of unusual rocky outcrops in the hope that they might be something valuable. He knew no metallurgy and less chemistry and had but very hazy ideas about geology. Yet the oldtimers who spent their lives playing a game with chance and against very heavy odds built up a large body of useful empirical knowledge and discovered many of the great mineral deposits of the world.

The searching for minerals was greatly improved when modern geology came into being with the work of the retired Scotch physician, James Hutton, a man of great intelligence and curiosity and a distressing lack of respect for authorities. He felt that the existing hazy ideas of the structure of the earth's crust were all wrong. He made careful observations of the numerous strata along the beaches of his seacoast farm. He ended up with a full realization of the importance of sedimentation in the forming of the character of the crust. He organized the history of sedimentary rocks on the basis of erosion, transportation, and deposition of materials. In 1785 he communicated his views to the then recently established Royal Society of Edinburgh in a paper entitled "Theory of Earth, or an Investigation of the Laws Observable in the Composition, Dissolution and Restoration of Land upon the Globe." The science of geology had its beginnings at that time. It gradually became known that certain types of minerals would be found in association with certain types of rock structure. This greatly limited the field of search and made the efforts of prospectors much more effective.

Though the work of most prospectors up to now has been largely a matter of surface scratching, the number of discoveries really has been amazing. The human animal, with prying insistence and undying optimism, has gone into every corner and noted everything that looked hopeful. A map of known deposits in the mineral-veined part of the country takes on the appearance of a face with smallpox.

Many of the hundreds of mineral deposits are undoubtedly deposits by courtesy only. Some of them will never amount to much in a commercial way. But these are mere skin indications. What then is underneath? What is there back under the ridges where no one has sampled? What would one find in a region such as this if he could investigate the entire area to a depth of 10,000 feet? The old-time prospector will never know, but new things are now on the program; new methods of prospecting are making it possible to investigate to great depths. We can hardly look forward to acres of diamonds under everyone's feet; but we may very well expect that new methods will bring new discoveries, perhaps important ones.

Within recent years the delicacy of physical instruments has been greatly increased, so now we find that the old hit-and-miss prospector has been partially replaced by the experts in geophysical prospecting. Five principal methods are now employed in this line of activity.

1] *Magnetic methods.* A body of iron ore, even deep in the ground, which contains appreciable quantities of magnetic iron oxide (Fe_3O_4) will cause a dipping magnetic needle to be pulled out of its normal position. This is caused by a vertical distribution of the magnetic field in the vicinity of the ore body. A sensitive magnetic needle which is confined to a single horizontal plane will show abnormal deviations in the apparent magnetic north if brought close to a hidden magnetic body. The magnet is a great aid in locating magnetic types of iron ore.

2] *Electrical methods.* If ground waters are in contact with sulfide ore bodies, voltaic cells are set up at various points in the ground. Careful measurement of gradients of electrical potential along the surface often aids in locating such bodies. Ores such as galena (lead sulfide) are better conductors of electricity than the average rock of the crust. Galena bodies are sometimes located by measurement of the electrical resistance of cross sections of the crust.

3] *Gravitational method.* The acceleration due to gravity (the physicist's friend, g) is appreciably increased in the vicinity of very dense ore bodies. A measurement of the time of swing of a pendulum affords a method of determining this constant, but that is usually too insensitive. The torsion balance, which balances the pull of the earth for two dense bodies through the torsional effect of a fine quartz thread, is usually employed. The measurements portray the presence of any large bodies under the surface which are of greater or less density than the average of surrounding rocks. This has been used extensively in locating the salt domes with which petroleum is associated.

4] *Seismic method.* The density and rigidity of a rock structure can be estimated from data of the speed of travel of shock waves through the solid strata. A small earthquake is caused by discharging a few hundred pounds of nitroglycerin under ground. (It not infrequently discharges prematurely.) The rate of travel of the shock is determined by timing the response of seismographs placed a considerable distance away. The waves travel at different speeds through different kinds of rock, and any unusual variation in character can thus be detected. Many of the salt domes which indicate possible petroleum deposits have been located in this manner.

5] *Radioactive methods.* Sensitive electroscopes, electrometers, or Geiger counters will detect even weakly radioactive minerals over a considerable distance. Radium or even uranium deposits can be located in this way, but the method is necessarily confined to the radioactive minerals.

The instruments of the physics laboratory have been a great aid in opening the storehouse of mineral wealth, but the field of geophysical prospecting is most decidedly one for carefully trained experts. The measurements have to be made very carefully and interpretation is by no means simple. But the methods are scientifically sound and have nothing in common with the superstition of the effectiveness of the old witch hazel or peach tree divining rod. Proper geophysical prospecting really produces results. Predictions often are not accurate; but the combination of geophysical methods and systematic searching has succeeded in taking a great deal of the chance—and a great deal of the supposed romance—out of prospecting.

Recovery Operations

After the prospector comes the miner, for no mineral is of any use as long as it stays in the ground. The method used for recovering the mineral from its earthen environment depends upon many conditions.

The traditional mine is a dark hole in the ground. A shaft, a sort of glorified well, is dug straight down to the level of the ore body. Then horizontal tunnels or drifts are cut out through the mineral body. The miner cuts and blasts his way through the vein of desirable material. The ore is loaded into small cars which are pushed by hand or drawn by small electric locomotives to the shaft and then it is hoisted to the surface. Unless the rock structure over the ore is unusually strong, it is necessary to support the roof of the tunnels after the ore has been removed. This calls for extensive bracing with heavy timbers. Often pillars of the mineral itself may be left in place to help support the roof. This obviously is a wasteful procedure, for a significant percentage of the material may be left under the ground in this way.

When the veins of mineral do not lie in horizontal strata or are badly broken, the ideal plan of a vertical shaft connected to numerous horizontal tunnels has to be modified to fit the case. Sloping shafts and numerous connecting tunnels at various levels have to be utilized.

There are attending facilities which must be installed in every mine. There must be considerable power available for transporting and hoisting the material. There must be adequate ventilation, particularly in coal mines where there is a gas hazard. The inevitable water must be pumped out. The problem of getting the water out of the English coal mines led to the demand for a satisfactory steam engine over a century and a half ago.

As long as horses were the only motive power, mines were shallow—at the most, a few hundred feet. When steam power came in, the water could be pumped out from almost any depth, and mining took on new life. The use of electric power has sent the mines still deeper. The deepest shaft in the world at the present time is down some 8380 feet in the Robinson Deep Mine near Johannesburg, South Africa. In this case mere ventilating fans are not sufficient, for the temperature in these deep workings is about 120° F.—with a humidity of 90 to 100 percent. Men cannot work under such conditions. Huge refrigerating plants have been installed to cool and dehumidify the air to make conditions tolerable for the workmen.

Every increase in depth brings more trouble and obviously more expense. Hence, only the richer ores can be mined at great depths, for the cost of the recovery must be less than the value of the mineral, but there is no definite limit to the depth to which mines might go. Modern methods and equipment can drive the mines to lower and lower levels, but every increase of a thousand feet calls for a greatly increased expenditure for operation. An ore has to be very valuable if it is to be feasible to recover it from depths greater than 5000 feet.

The Hull Rust Mine at Hibbing, Minnesota, looks like part of the Grand Canyon of the Colorado moved north. No burrowing like moles here. Nature obligingly put down only about fifty feet of soil and debris on top of the great beds of iron ore. The overburden was stripped off with steam shovels, and other power shovels began digging out the great hole by loading the ore directly into cars. The general appearance of the mine is that of an inverted pyramid with the base at the top. The sides are terraced in wide shelves which descend hundreds of feet in ever closing circles to the bottom of the deposit. Another great operation is in the open-pit copper mines at Bingham, Utah. Many of the coal seams in Indiana and Illinois are covered with only a few feet of debris, and open-pit mining is widely utilized there, too. The largest power shovels in the world (one of them takes up thirty-two cubic yards of dirt in one scoop) are used for stripping off the overburden. Some of the lignite deposits of Germany are worked by open-pit methods.

Presumably almost any mine could be operated by open-pit methods, but if the overburden is large the problem of its complete removal becomes enormous. Finally a point is reached where it is more economical to sink a shaft and let the overburden stay in place. It is seldom feasible to use open-pit methods with overlying debris more than fifty feet thick.

When the mineral to be recovered is a liquid, the mining process consists of drilling a hole in the ground—in the right place—and putting a pump in operation. The largest scale recovery operation of this type is in the petroleum industry.

Sometimes nature has stored up gas in the tight pockets of rock overlying an oil deposit, and the pump can be dispensed with during the first part of the well's history. The accumulations of millions of years' deposits of organic decomposition sometimes build up a gas pressure greater than a thousand pounds per square inch. The earth then has a local condition resembling that of a severely abscessed tooth; and, when the drill tool breaks through the overlying solid structure, the earth's gas accumulation tries to correct the situation by forcing the oil up through the newly made relief passage with a rush and a roar. This is the beginning of the "gusher," the only kind of oil well that often gets into the public press. Most of the wells are more sedate and less exciting; the oil often has to be pumped to the surface. As the sands or porous rocks in the immediate vicinity of the well casing are depleted, more oil seeps in from the surrounding structure to be picked up by the sucking action of the pump.

Gas wells are not pumped. If the pressure is so low that the gas cannot flow out through the open hole, then there isn't enough gas to be worth recovering. In a great many cases the gas has to be compressed after it leaves the well in order to send it through pipe lines with a reasonable speed.

The world's record depth for a petroleum well at the time of this writing (1938) is 15,004 feet, almost twice as deep as the deepest mine. Wells of such depth could not be possible without power equipment and rotary drills; but, even with the finest equipment, drilling 15,000 feet straight down through solid rock is a difficult and expensive task, but deep wells are not entirely new. The Chinese drilled wells over 3000 feet deep to obtain salt water for salt making in the Middle Ages. Such a well, drilled without mechanical motive power and with only the crudest tools, was a far greater accomplishment than the 15,000-foot well of today. Ingenuity is no monopoly of the modern world.

There are other valuable liquid minerals besides petroleum. Brine has been pumped from the ground for salt making for thousands of years. Over two thirds of the salt recovered in this country is obtained by evaporating brines from wells. In recent years the impurities in some of the brines have become the base of a large chemical industry. The brine wells of Midland, Michigan, are now the source not only of salt (sodium chloride) but of bromine, Epsom salts, and metallic magnesium. There are several brine wells in California that are used to produce iodine.

Sulfur is another mineral that often comes from wells. The most important sulfur deposits in the world occur along the Gulf coast of Texas. The first sulfur recovered by wells came from Louisiana but that deposit has been almost exhausted. This sulfur in Louisiana and Texas was deposited apparently in the free state by the oxidizing action of certain bacteria on hydrogen sulfide gas, H_2S. The Frasch process for sulfur recovery uses a well with a double casing. A mixture of steam and water at 250 pounds pressure (175° C.) is pumped down the outer annular space. This melts the sulfur in the vicinity of the well, which then flows into the casing and is brought up the central casing by means of air. This simple process, which was hooted at when it was proposed, has completely revolutionized sulfur recovery. It not only is an inexpensive, rapid method, but by the process of selective melting purifies the sulfur before it leaves the bed.

There is still another kind of mineral deposit, which does not require even so much as a well; it is simply a case of come and get it. Many of the brine lakes which are scattered over the surface of the earth contain compounds that are worth recovering. The Dead Sea lived on its history for several thousand years, but now it is doing its turn as a source of crude salt for the recovery of potassium salts for fertilizer. The brine pumped from Searles Lake in California supplies most of the borax used in this country and a significant part of the potash.

Then there is the most nearly limitless mineral source of all—the ocean. Sea water has been used occasionally as a source of salt for seasoning for thousands of years, but only recently has it been used as a source of material for chemical treatment and recovery operations. Liquid bromine is now being extracted from sea water at a chemical plant on Cape Fear just south of Wilmington, North Carolina. The plant started operations January 10, 1934. That was a landmark in chemical industry. The possibilities of recovery of elements from sea water have by no means been exhausted. More will be heard of this in the future.

In the same sense that the sea is a mine, so is the atmosphere. The air is now serving as an absolutely boundless mine of free nitrogen, which is the chemical raw material for making ammonia and hence nitric acid. It is available merely for the taking. No legislature has found a way to impose a tax on it yet.

Beneficiation of Ores

There are many steps on the mineral stair before final utilization is reached. First, there were the various steps in the segregation of compounds to make ores—processes which had no connection with human existence. Then the prospector had to find the mineral deposits. Then the miner, or the driller, or the pump operator had to remove the mineral from the ground, the well, or the ocean. Still the succession was not complete. In most cases some operations had to be performed to bring the desirable material to a higher concentration than it had in the natural ore, before it could successfully be processed.

There are very few of the solid minerals which do not go through some form of beneficiation to improve their quality before they reach the smelter or the market. A catchall term, "ore dressing," is often used to designate whatever is done to an ore to improve its quality, though some sticklers for usage may contend that the term refers only to the mechanical operations which may be employed.

The most obvious way to get ore of good quality is to pick up only the better pieces. This technique probably was followed in the gold mines of the Pharaohs and of King Solomon. Where the desired material is easily recognized and not too finely divided, and where labor is inexpensive, this is one of the more effective ways of concentrating ores. It is followed more or less today. Miners often do a certain amount of selection of the material sent to the surface, and hand picking from belt conveyors is still used almost universally to segregate slate from coal.

Every mining industry has a variety of methods and apparatus for making little rocks out of big ones. Glass-hard metal jaws seize large rocks and crush them to bits. Gyratory crushers take rocks as large as a man's head, that have withstood the elements of weathering for thousands of years, and squeeze them into little nodules in an instant. There are hammer mills, ball mills, rod mills, stamp mills, driven by large motors, that pound and smash away at their tasks in deafening and dusty surroundings. The object of all of these activities is to break the rocky material down so that each desired piece of ore stands by itself, is not attached to impurities or gangue. Sometimes, as in the case of certain iron ores, it is necessary only to crush to five-inch pieces, for iron comes in large pieces. Gold ore comes from its stamp mills ground to an almost impalpable powder, for gold usually occurs as minute flakes in the rock. After grinding comes the separation of the desirable from the undesirable.

In some cases there is a distinct difference in the friability of an ore and its attending impurities. If the ore is soft and friable and the gangue (impurities) is hard and tough, a crushing operation will break up the ore into smaller pieces than the gangue. In such a case, rocks after crushing may be separated by the use of screens. The small-sized material will be rich in the desired ore, the large-sized material will contain but little.

If a mixture of mineral particles of various sizes and densities is allowed to fall through water from a given height, all pieces starting at one instant, the individual particles will form layers according to their densities and size. If all particles are the same size, the densest particles will form a layer on the bottom, with the least dense on top, the intermediate material graded in between. If the sizes are mixed, some large-sized particles of low density may be mixed with some small particles of high density; but all of the particles in any one layer have "equal fall" periods. By carefully controlling size, it is possible to make very good separations on the basis of density alone. The pieces of equipment used for gravity separation are as varied and as strange as their names: jigs, tables, vanners, sluice boxes, and classifiers. It is impossible to give one-sentence explanations of the operations of each of these.

The "gravitational" methods of ore concentration depend on the differences in rate of settling of particles of different densities in water; but, if given time enough, all particles settle. If particles of ore and particles of undesirables have distinct differences in density, it seems quite obvious that the simplest and most efficient way to separate them is to put them into a liquid with a density intermediate between the two solids. The heavy one will sink, the light one will float; the separation is perfect. All you have to do is to find the liquid. The liquid must fulfill several requirements. It must be relatively inexpensive, it must not be volatile, it must not be absorbed to any large degree by the solids, and above all it must have the correct density. Nature made things surprisingly difficult for man on this score. Among the solids there are rocks and metals with specific gravities well spread from 0.534 (lithium) to as high as 22.48 (osmium). Among the liquids there are water with a specific gravity of 1.00 and mercury at 13.6. But of the materials that are liquid at ordinary temperatures, there are very few that have intermediate specific gravities. There isn't much in between to be used for float and sink operations for the separation of minerals. Liquid bromine has a specific gravity of 3.12 but it is a volatile, highly obnoxious, and expensive element. It is unusable on all counts.

The organic chemists, having small regard for nature's limitations, have synthesized several liquids somewhat heavier than water. For instance, there are chloroform ($CHCl_3$) specific gravity 1.50, carbon tetrachloride (CCl_4) specific gravity 1.58, pentachlorethane (C_2HCl_5) specific gravity 1.83, and tetrabromethane ($C_2H_2Br_4$) specific gravity 2.97. The latter two have proved to be highly successful as media for the separation of coal (it floats) from slate (it sinks). Prices of the organic liquids have been lowered and methods of minimizing losses have been improved so that the float and sink method stands an excellent chance of replacing the eyes and hands of sorting workmen in the old coal breaker. Pilot-plant operations at Shenandoah, Pennsylvania, have shown that the method is inexpensive, requires only a small amount of equipment, is infinitely cleaner than the old hand picking, and gets very much better results. "Float and sink" has a great future in the coal business, for improved efficiency of separation is badly needed. If someone will devise suitable liquids of intermediate specific gravity for use in the metal ore fields, there is a great future there, too.

One of the simplest ways of separating iron filings from dirt is to pass a small magnet over a thin layer of the mixture. The iron filings jump to the magnet as if it were a long-lost friend. Magnetic iron oxide (Fe_3O_4) can easily be separated from undesirable material in the same way. The separating is usually done by means of a conveyor belt with electromagnets placed underneath to hold the iron oxide in place longer than the nonmagnetic material. Any iron ore can be rendered magnetic by roasting (heating in the presence of air) at a few hundred degrees C. Hence the magnetic separation can be employed for any iron ore.

Iron is sometimes an impurity rather than the desired material. Zinc ores are frequently benefited by the removal of pieces of iron ore.

One of the few types of experiments with which the ancient Greeks would dirty their hands was electrostatics. A piece of amber rubbed with cat's fur attracts a pith ball. The pith ball will cling to the electrically charged amber almost indefinitely in a dry atmosphere. A piece of highly conductive material, however, will cling to the charged amber only momentarily, for the electrons which make up its charge are quickly dissipated to the charging body. This difference in the length of time of retention of charges is the basis of the electrostatic method of separating minerals from impurities. The variable property in this case is that of electrical conductivity. Finely divided ore is put onto the top of a metallic roller which is one of the electrodes of the electrical system. A high potential electrode rests above the roller and sprays the mineral layer with electrons—a high potential brush discharge. The minerals, which are relatively good conductors (lead sulfide, for instance), lose their charge to the roller immediately and fall off into a hopper. The poorly conducting particles (the impurities) cling to the roller and are scraped off on the opposite side into another hopper.

We have discussed separations of valuable from undesirable materials on the basis of differences in friability, specific gravity, magnetic susceptibility, electric conductivity; and now comes a separation based on differences in wettability. "Wettability" is one of those catchall phrases used to designate something that is not completely understood. When a liquid spreads over a solid in a thin and tenacious layer, the solid is easily wettable. The system, hands–dirt–soap–water, offers an illustration. When the system, hands–dirt, is brought into contact with the system, soap–water, with a certain amount of agitation, the dirt becomes thoroughly wetted with thin layers of soap in water. There is a change in the interfacial tensions, the dirt clings tenaciously to the soap–water phase and leaves the hands. The same principles of this cleaning process are applied on a large scale to the concentration of minerals.

There is an unverified story to the effect that a gold miner's wife told her husband that she would show the men a thing or two about the process of mineral concentration. If soap could pull dirt out of clothes, it ought to pull gold away from pieces of rock. She tried it on some gold-bearing sand and found that the soapy froth carried a relatively large amount of gold in it. By scraping off the froth she concentrated the gold. The story's authenticity is a little doubtful because soapsuds are not very effective in separating gold from gangue. Nevertheless, someone worked a major revolution in metallurgy a generation ago when he started the investigation of the possibilities of "flotation" through the floating of desirable ore particles on the froth of a water suspension of the finely divided raw ore.

In this process the entire ore is ground to very fine dust, in ball mills or in stamp mills, and is mixed with water to form a slime or sludge. A small amount of oily material, such as oil of eucalyptus, coal-tar creosote, or ethyl xanthate, is introduced into the water. If the correct agent has been added to the water, the oily material dashes to the surface of the desirable mineral particles and forms a very thin, perhaps mono-molecular, layer on them. The vat of sludge is then agitated with air. A froth of air bubbles rises to the top and the oily part of the suds carries the ore particles upward with it. The gangue materials are not wetted by the added organic agent, so they stay behind in the bottom of the vat. Strangely, it is usually the heavier solid particles that are "floated." They "float" because the air-filled froth which contains the finely divided mineral in the oily films is lighter than water. The froth is skimmed off with paddles, like skimming the foam off a glass of beer, and the mineral is obtained in practically a pure condition.

The treatment of each mineral and each ore presents a research project of its own. Every system has distinct behavior problems that can be solved only by extensive investigation. Flotation was first used commercially in this country in 1912 when the Butte and Superior Mining Company at Butte, Montana, installed a plant to handle the slime of a lead-zinc ore. The process is most successful on sulfide ores, such as those of copper, lead, and zinc; but it has found some use in gold and silver recovery. Research work has been carried on for the separation of almost every mineral from its impurities. It appears that a selective wetting agent can be found for practically any solid substance and that it can be floated in a froth. Finely divided coal has been successfully separated from slate in this way and iron ore can be floated away from siliceous impurities. These last two have not been applied commercially as yet, for some of the other mechanical methods are less expensive.

Aluminum

The foregoing is a bare outline of some of the most successful mechanical and electrical tricks used to concentrate, or purify, minerals; but chemistry makes large contributions, too.

The production of pure aluminum calls for the use of pure aluminum oxide. The mineral now universally used for aluminum production is bauxite ($Al_2O_3 \cdot 2H_2O$). But bauxite always contains impurities, such as iron and silicon compounds. These must be separated from the alumina (Al_2O_3) with which they are intimately associated, and only chemical methods have been successful.

In the most used method the aluminum oxide is converted to a soluble aluminate by heating with a strong solution of sodium hydroxide. The aluminate is soluble and thus separates from the impurities which are insoluble in an alkaline solution. The aluminum is then precipitated as aluminum hydroxide, $Al(OH)_3$, by cooling the solution. Impurities in the solution are not precipitated. The aluminum hydroxide is decomposed to alumina (Al_2O_3) by heating.

Gold

Gold concentration and recovery have had frequent contact with chemistry. Metallic gold has a strong affinity for mercury, forming a silvery amalgam as anyone can determine by rubbing a drop of mercury on a gold ring. Gold is secured from many "free-milling" ores (those which break up easily) by an amalgamation process. The finely ground ore is brought into contact with liquid mercury. The fine ore particles are dissolved in (amalgamated with) the mercury. The liquid amalgam when saturated with gold is drawn off and placed in a retort. The mercury is distilled off and condensed for reuse. The gold remains behind.

The cyanide process for gold recovery was invented in 1891 and has now almost superseded the other methods of recovery of finely divided particles of gold. Finely ground gold ore is thoroughly washed with a dilute solution of potassium cyanide. Gold cyanide is formed, which is soluble in water. The rest of the solid matter of the ore is insoluble. The solution containing the dissolved gold is filtered off and the gold is precipitated by means of zinc shavings or by electrolysis.

Bromine

Bromine (used in making photographic emulsions, headache powders, antiknock gasolines) can be extracted from sea water where it is present in minute proportions. The water is acidified to a certain, carefully controlled point (pH of 3·5 to 4) and gaseous chlorine is pumped into the water to replace the bromine. The bromine is then stripped out of the water solution by means of an air stream.

What is an Ore?

This discussion of the processes of ore formation and methods of recovery has been merely leading up to a discussion of the question: What is an ore? There is no definite answer. The miner's definition of an ore is a mineral that can be recovered and worked at a profit. The items which must enter into consideration are the value of the final product (in dollars); the concentration of the desired constituent in the deposit (solid or liquid); the accessibility of the deposit; the number, character, and chemical and physical conditions of the undesirable impurities.

A deposit of pure copper 25,000 feet under the surface would not be a usable ore at the present time, for we cannot yet operate mines at a depth of 25,000 feet. Pure iron oxide on top of an inaccessible mountain in the middle of a wilderness would not be a usable ore, for operators could not get to it nor could it be taken out.

The concentration of the material is one of the chief considerations. As rich deposits have been depleted, new techniques have been devised to handle poorer material. In the old days of the West, gold was mined from quartz veins by grinding the material and separating the heavy particles of gold in sluice boxes (gravity separation). The dumps of spent ore grew and grew and sullied the landscape; but these dumps still contained considerable gold. Then came the amalgamation process of gold recovery which would remove about 60 percent of the metal. This new technique changed the old ore dumps into gold mines. The once useless material was reground and the gold recovered by means of mercury. The tailings of these mills were dumped to choke the valleys again. After a time the cyanide process came in and the dumps from the amalgamation mills were again valuable ores for the new process, which could remove about 98 percent of the gold. Again, the old ore dumps were reused and gold was recovered at a profit.

The rich copper ores of the West were seriously depleted by 1911 when the flotation process came into use. When the new process of concentration was shown to be successful for copper ore concentration, old mines were reopened, former low-grade deposits became valuable.

A pile of dirt today may be transformed into a usable ore tomorrow by the discoveries of some research worker in a distant laboratory. Mineral resources are only half determined by the deposits which nature has handed to us. The other half is the ingenuity of the chemist and the engineer.

A large part of the story of the individual sciences can be told in terms of the tools, instruments, and machines they employ. The history of astronomy is concerned not only with the theories of Copernicus and Newton but with the observations and proofs provided by the telescope, the radio telescope, and the spectroheliograph. The science of bacteriology was founded in the seventeenth century when a Dutchman named Leeuwenhoek observed "very small living creatures" through lenses he had ground. The modern physicist uses giant cyclotrons to shatter atoms for research. The geologist measures the distance, direction, and force of earthquakes with the seismograph. There is hardly a discipline in which the camera, like the written word, does not preserve knowledge.

The engineer, as we have seen, first relied on relatively simple tools, like the lever, the wheel, and the inclined plane. As they grew in complexity, such tools, utilizing new scientific principles, developed into machines like the steam engine, the turbine, and the dynamo. Today, machines exist which perform many of the functions of the human brain with so much greater speed and accuracy that they pose problems of great sociological and economic importance. The dilemma is not new. Throughout history, refinements in tools and techniques have proved long-term blessings but have often meant misery for large sections of the population. The word "saboteur" has its origins in one revolt of man against the machine. In his book, Automation, Samuel Lilley mentions Karel Capek's play R.U.R. as a dramatization of the hatred and fear of human beings for the "robots" which threaten to take their place.

How much justification is there for this deep-rooted suspicion of the machine? Can automation ever dispense with human supervision? What type of workman is apt to be displaced? Will the resultant benefits justify the technological unemployment involved and what can be done to cushion the blow? The selection reprinted below opens vistas of a society in which far less effort will be spent on production and far more on creativity and enjoyment.

TOOLS AND AUTOMATION

SAMUEL LILLEY

THE FIRST QUESTION that needs answering is: What are we talking about? What do we mean by this rather ugly new word, Automation?

To get started, let us be content to define automation provisionally as:

> The introduction or use of highly automatic machinery or processes which largely eliminate human labor *and detailed human control.*

It is the italicised phrase that is important and distinguishes automation from mere mechanization. In mechanization the machine takes over the work, but the human operator is still required to exercise detailed control over its operations—to observe at every stage just what the machine has done and then to operate controls which give it instructions as to what it must do next. Automation abolishes that sort of control. In the ideal case (though naturally practice often falls short of this) the machine completely controls its own actions; it carries through a manufacturing process from beginning to end without any direct help from human beings; and the human worker is required only to see that the machine is kept in good running order.

If we are to appreciate the impact of automation today, we shall first need some understanding of engineering production methods in the preautomation era and of how they grew up. It was about 1800 that Henry Maudslay transformed the old woodcutting lathe and the refined watchmaker's lathe of the eighteenth century into the robust and accurate metal-cutting lathe which is the fundamental tool of engineering. And it was about the same time that the first steps were taken toward a new way of organizing the production processes as a whole, which has been the main road of development ever since.

The new method was not mass production as such, but production on the principle of interchangeability, which facilitated the extension of mass production to quite new fields. The new method of interchangeable manufacture was based on the principle that all the parts, as produced, must conform to standard measurements (within permitted variations, called "tolerances") so that an unskilled assembler can take at random any set of them, knowing in advance that they will fit one another, and can build them up into the completed article without further manipulation. The highly skilled fitter is replaced by the unskilled or sometimes semi-skilled assembler.

Maudslay was associated with one early enterprise on these lines—that of Marc I. Brunel and Samuel Bentham (brother of the more famous Jeremy) for making pulley blocks in 1808. Every block consists of several parts. A single frigate needed 1,500 of them, and the Admiralty at that time used 100,000 yearly.

Brunel and Bentham conceived the idea of producing blocks by interchangeable manufacture, and worked out the general scheme of the machine to be used, which was put into execution by Maudslay. The project was designed to use unskilled labor—indeed, convict labor, for it was connected with Jeremy Bentham's prison reform schemes. And in fact ten unskilled men did work that would previously have needed 110 skilled. They produced 130,000 blocks a year, equivalent to the previous output of the six largest dockyards.

The scene shifts to America, where Eli Whitney started making muskets by interchangeable methods in 1800, and Simeon North did the same for pistols a little later. At first they used water-powered machinery for rough processes, but the accurate finishing was done by specialized workmanship and by filing and drilling in standard jigs, not by special machinery.

They soon, however, began to feel the need for machinery specially designed for the purpose of turning out components at high speed and with unvarying accuracy. In 1818 Whitney produced the first milling machine, and Blanchard the "gun-stocking lathe" for turning irregular shapes by accurately following a pattern. The turret or capstan lathe, in which several tools set in a revolving turret can be brought successively to bear on the work as the operator manipulates the controls, appeared about the middle of the century. Through the addition of various automatic connections between the working parts, it gradually developed from the 'sixties onwards into the automatic lathe. The universal milling machine appeared in 1861. And so the development goes on, through the multispindle automatic lathe, the cylindrical grinder, the vertical turret lathe and many others. Similar developments were taking place in such operations as drop-forging and die-stamping. And before the end of the nineteenth century most of the basic range of machine tools had appeared.

It was the small arms industry that established interchangeable manufacture. And for the first fifty years or more, most of the new developments in machine tools were made in connection with that industry. But gradually other industries took up the method—again almost exclusively in the U.S.A. It was applied to the making of wooden clocks by 1809, and a little later to brass clocks; by 1855 these were being produced at the rate of 400,000 a year at 50 cents apiece. In the 'fifties interchangeable manufacture was fundamental to the widespread success of the sewing machine. And for a different reason it was equally fundamental to the widespread use of agricultural machinery in the 'sixties. This time the point was to solve the problem of maintenance in rural areas where skilled labor was not available. The farmer ordered replacement parts by catalogue number, knowing that they would slip easily into place. Interchangeable manufacture was also applied to the mass production of typewriters in the 'eighties, and of bicycles—several hundred thousand a year—in the 'nineties. And so by the end of the century, just as the motorcar was coming along to give it renewed impetus, the method had become the established way of producing engineered goods in large quantities.

It was the car industry that established that other major characteristic of modern practice, the assembly line. But even that has a prehistory that goes back well into the last century. In 1833 the British Admiralty set up a remarkable bakery for ship's biscuits. It used machinery for kneading and other operations, but the chief novelty was that the work passed from one workman to another on trays running on power-driven roller conveyors. Again, we find Johann Georg Bodmer, a Swiss working in England, building a factory about 1840 for making textile machinery, with special attention to the arrangement of the machines so as to minimize transport, and with such devices as overhead cranes and conveyor belts.

But once more the scene shifts to the U.S.A., and the first real appearance of the assembly line was at Cincinnati in the 'sixties—even though it was applied not to the assembling of engineering products but to the *dis*assembling of pigs, to cutting them up for canning. The carcasses were carried on overhead conveyors past a series of workers, each of whom had the job of making a single cut or removing a single piece. These methods were developed further in Chicago, where they reached such a pitch of efficiency and economy that—so it is said—they even used the pig's dying squeal to work the factory hooter. Something approaching the assembly line was used in building railway freight cars in the 'nineties—with the chassis towed along rails between workmen who each did only a few specialized operations. But the full advent of the modern assembly line was a consequence of the mass-production car industry.

The early cars were built as individuals. But about 1902 some firms started moderate scale production using interchangeable methods. Olds produced 2,500 cars in 1902 and 5,000 in 1904. Ford's production passed 10,000 in 1909. The industry used all the machine tools that earlier mass production had created and added many more—precision gear-cutters, precision grinders of many types, a great variety of presses, stamps, die-casters, drilling and tapping machines, and much else. And then in 1913–1914 Ford introduced the assembly line. He first tried a conveyor-belt system for assembling flywheel magnetos, and succeeded in reducing the assembly time from twenty minutes to five. This led to applications in the assembly of other small parts, and shortly after to experiments concerning the assembly of the whole car. In the first of these the chassis was towed along a 250-foot length of street beside the factory past stockpiles of parts placed at regular intervals, and this reduced chassis assembly time from fourteen hours to six. The principle of the method was now established, and it was not long before it had taken on the more familiar conveyor form. Ford himself summarised the procedure in these words: "The Ford shop's assembling practice is to place the most suitable component on elevated ways or rails and to carry it past successive groups of workmen, who fix the various components to the principal component, until the assembling is completed."

There have been many changes and improvements since then. Many new machines have been introduced, the most important perhaps being the centerless grinder. Systems of marshaling the components for the assembly line have been steadily improved. And the machining also has become more and more organized on a flow, rather than a batch, basis—the work passing progressively through whole lines of machines. But in essentials, the practice that emerged in the Ford works about 1914 is the engineering production practice of today —except in so far as it has very recently been modified by automation.

The essentials of this practice are:

(*a*) Machining on machines so designed that, once they have been correctly set, they will turn out work repetitively to the required measurements in the hands of semiskilled or even unskilled operators.

(*b*) Conveyor-belt assembly, in which a main component flows continuously past many workers, who attach the other components to it.

Experience has shown that this system of production is a very efficient one.

But the system is not free from disadvantages. The degree of division of labor continually increases. From this there results a progressive degradation in the amount of skill required from the factory hand. The ideal worker becomes a brainless automaton capable of repeating a few simple movements with very great speed and dexterity, but not required to exercise any judgment.

The problem—as so often in history—leads on to its own solution. Once a production operation has been reduced to a form in which the human worker has merely to act like a machine, then it is not technically difficult to take the further step of substituting a steel and oil machine for a flesh and blood one, the step of making the process entirely automatic.

That is what is beginning to happen. Actually the motive that has led to the strong trend towards automation in the last few years has not been the improvement of working conditions. The real driving force has been management's wish to reduce labor costs.

Automation did not become important till after the Second World War. But it has a prehistory stretching back to the 1920's. Indeed, we could stretch its prehistory a good deal further if we agreed to include such items as the automatic flour mill set up by Oliver Evans near Philadelphia in 1784–1785. But, though it certainly worked, this can perhaps be dismissed for our purpose as an historical curiosity.

A more serious start was made in 1923–1924, when Morris Motors installed transfer machines for machining cylinder blocks and other large components. The cylinder block machine, 181 feet long, did 53 operations at the rate of 15 blocks an hour, and needed only 21 operators. Because these machines had to rely entirely on mechanical controls, they proved too unreliable and had to be abandoned.

In 1928, A. O. Smith and Co. of Milwaukee set up a very remarkable plant for making car frames entirely automatically. It began with steel strip, which passed first through an inspection station, where it was checked and straightened. Thereafter, as the work passed automatically from station to station, the strips were cut, bent, punched and pressed in the various forms needed for different parts of the chassis. Still automatically the parts were brought together and riveted, and finally the completed frames were brushed and cleaned ready for painting. Run by a staff of 120, mostly supervisory and maintenance, it produced about 10,000 frames a day, so that each frame was the product of only sixteen man-minutes of work.

In the U.S.S.R. parallel developments were taking place. In 1939, Foreman Inochkin of the Stalingrad Tractor works linked five lathes together by automatic conveyors to perform a sequence of ten operations on caterpillar roller bushings. And about the same time a remarkable automatic iron-casting machine came into use at Klimovsk. Using permanent metal molds on a rotary caster, this plant had an output of 10,000 pieces in a two-shift day.

Though these machines are obviously direct predecessors of the automated lines of today, the biggest contribution of the inter-war period to the preparation for widespread automation was probably in the development of electronics.

While it is probably true that even today the greater part of automated machinery in use manages to do without electronics—using mechanical or electromechanical devices instead—it is equally clear that the future of automation lies very largely with electronic control.

The reason for this is that only in very simple cases can a mechanism be set in advance to complete a given job without need for adjustment in the course of the work. It is easy to verify this point by setting oneself a simple task such as that of picking up a pen from a table at the other side of the room. I try to preset myself to achieve this object—that is to say, I look to see where the pen is, and then, shutting my eyes, I try to walk across the room and pick it up. Almost always I am very wide of the mark. Contrast that with my normal procedure. After preliminary observation of the pen's position, I start to walk towards it with my eyes open; when I get near enough I move my hand in the correct direction; and finally, as my hand comes into position, I bend my fingers to grasp the pen. But all the time I am making use of my eyes and brain. In the early stages my eyes send a succession of messages to my brain, saying: "The table is now so many feet and inches from you in such-and-such a direction." My brain considers these messages and transmits instructions to my legs concerning such points as the length of the next stride they are to take. Soon comes a message from the eyes reporting that I am within arm's length of the pen. Here my brain tells my legs to stop moving and starts sending instructions to my arm muscles to move in such-and-such directions. As this movement goes on, my eyes make further reports concerning how close my hand is getting to the pen, and from moment to moment my brain sends further orders to the muscles to modify the direction and speed of my arm movements. Soon comes the message from the eyes to say that my hand is in position to grasp the pen, and so my brain starts telling my fingers to bend—and so on till the operation has been completed.

This is the process known as feedback. Information from my eyes about how well my limbs are getting on with the job is fed back through the "telephone exchange" in my brain and used to modify the action of my muscles from moment to moment.

Similar conditions apply to the work of many machines. In all but the simplest cases the operator must watch the progress of the work and take action from time to time to modify the machine's behavior. And if the process is to be made automatic, then it is necessary to find a mechanical or electrical feedback mechanism to take the place of the operator.

The principle of feedback control of machinery is far from new. A steam engine is required to run at constant speed. That cannot be ensured by a fixed throttle setting, since variations in the load or in the activity of the stoker will affect the speed. A man could be set to watch the engine, with his hand on the throttle control ready to open it a little if the engine slows down or close it when the revolutions increase. But the centrifugal governor which James Watt invented in 1788 does the job much more efficiently. A somewhat similar device is used in a ship's steering gear.

These examples show that feedback can be achieved by purely mechanical means. But only rather simple feedback jobs can be dealt with mechanically. For most types of automation that are common at present, electromagnetic methods are good enough. But for really advanced automation electronic control will often be required. The main thing that gives electronics the advantage is that the speed at which information is fed back is often of crucial importance. If there is too much delay, the machine will "hunt"—that is, it will oscillate backwards and forwards on either side of the required behavior, but will never be dead on. That again can be easily demonstrated by a variation of the experiment of picking up the pen. I arrange to put a considerable delay into the feedback mechanism by keeping my eyes shut and getting my information from a friend who tells me just how my hand is placed in relation to the pen. It will be found that my hand moves hither and thither about the pen for quite a while before I succeed in grasping it. Another convenient way of introducing a delay in the feedback is by the immoderate consumption of alcohol—and anybody watching a drunken man trying to pick up that pen will see plenty of examples of hunting. Now electromagnetic relays work in tenths, or at the best hundredths of a second, while electronic valves can cope with millionths. That is why electronic control methods are essential in cases where really fast feedback is needed.

Then again, any feedback mechanism necessarily includes a part which corresponds to the human sense organ, a part which observes what has happened up to date. Usually the observation takes the form of a measurement—the size of a component that is being machined, the temperature of a vessel in which a chemical reaction is taking place. Some observing and measuring jobs can be done mechanically —Watt's governor measures speed; mechanical "fingers" can "feel" the size of a piece being worked in a lathe or a grinder. But again, electronic devices have great advantages —the photo-electric cell, which produces a small electric current when light falls on it, can perform the functions of an eye; other electrical and electronic devices can estimate chemical composition, or do jobs analogous to that of the railway wheel-tapper who tests the soundness of carriage wheels by the noise they produce when hit with a hammer.

Most of the main electronic devices were invented before the First World War, but it was only in the 'thirties and after that their potentialities as production control devices were seriously explored. Then photo-cells came to be used for such jobs as grading and sorting rice, beans, and cigars by color (they would have taken over from that chap who disliked the responsibility of cherry sorting); detecting and rejecting unlabeled tins coming off an automatic labeling machine; and removing metal bars from a furnace at a specified temperature (measured by the light they emit). All these are simple cases of feedback—the action of the machine is varied according to what the photo-cell "observes." There were also some applications of photo-cells to inspect articles coming off machine tools—for example, in an American unit for inspecting camshafts, which enabled four men to do the work of a former 18—though, so far as I know, it was not until the late

1940's that any inspection device was linked with a feedback circuit for adjusting the setting of the machine.

Another way of using photo-cells was exemplified in the Russian development in 1940 of a lathe controlled by a photo-cell which reads a blueprint and directs the actions of the tools (on one or many machines) accordingly. A similar application, made in the U.S.A. during the war, was to the control of an oxyacetylene cutting machine by a photo-cell reading a scale drawing. These were important anticipations of what are now called "program-controlled" machines.

Wartime problems like that of keeping an anti-aircraft gun trained on its moving target led to rapid developments in electronic feedback mechanisms, while radar work produced extremely rapid progress in electronics generally. The war, too, was the time when work began on the development of electronic computers ("electronic brains") whose role in automation is becoming more and more important. As a result of all this, we in the postwar era have at our disposal a range of electronic equipment that is apparently capable of dealing with most of the control and feedback problems that automation will pose.

But despite the impression created by many popular articles in the Press—most automated machinery in use at present does *not* employ electronic controls, but is sufficiently well served by the cruder electromagnetic relay.

Let us suppose that we have to machine an Austin A40 cylinder block. The foundry supplies a rough casting, but to turn this into the finished article several hundred operations will have to be done. A great many small holes for bolts or for oil channels must be drilled. The cylinder bores and the holes to take the camshaft and crankshaft will probably need three operations each—a preparatory drilling, a semifinishing operation and the finish boring. Various faces will have to be milled smooth. And so on.

Until a few years ago, a series of jobs like this would be done on a long line of individual machines. Each machine might do several operations simultaneously—drill a number of parallel holes, for example. Each machine might, as a separate unit, be substantially automatic—in a driller, for instance, the operator might have to place the block in position and move a lever to start the motors; then the drills would begin revolving, advance slowly until the holes were drilled to the correct depth, retract to their original positions, and stop; here the operator would take over again to unload the piece on to the conveyor that would carry it to the next machine.

We are now asked to find a method of substantially reducing the cost of this work. Individually, the machines have already come very near to perfection. There is little possibility of making them work faster, for example, or of saving significantly on the power they consume. But well over a third of the cost lies in the wages of the operators. And, looked at from a purely engineering point of view, the operators contribute very little to the job. They do a few simple loading and unloading movements, and one or two manipulations of levers and switches, which look quite trivial compared with the work of the machine in removing large quantities of unwanted metal and doing it to a very high degree of accuracy. Would it be possible to do without these operators—to arrange for the machines themselves to take over the few simple jobs which they still leave to their attendants?

One could, perhaps, arrange for mechanical hands to load the blocks from conveyor to machine and unload them from machine to conveyor, and connect these mechanisms to the motor switches, so that the power is switched on and off as required. That is sometimes done. But for most purposes it would be a clumsy compromise, and it is usually better to redesign the machine completely as an integrated unit. What we now have is a Transfer Machine—the bread-and-butter of automation. Instead of a series of separate machines, there is a series of stations, each of which does the work of one of the old machines. The cylinder blocks lie in a channel that passes through the center of these stations, and they are tranferred from one station to another automatically.

Let us follow through a complete cycle of operations. If a cycle has just finished, there will be a partly worked block in every working station, and a finished one in the unloading station at the end. An operator removes this finished block, puts it on the conveyor, and presses a button to signal "All clear." Meanwhile another operator places an unworked block in the empty loading station and also presses a button to indicate that he has done so. The transfer mechanism comes into action and moves every block one station forward. Then the various tools in all the stations advance simultaneously to do their various jobs. When they have finished, they automatically retract and stop, and the cycle is finished. Every block in the line has had one further set of operations done on it; there is another block ready for unloading and an empty loading station ready to receive the next block starting on its journey.

In some cases it may be necessary to turn the work upside down, so that the tools can be brought to bear on the bottom, or to turn it through a right-angle horizontally. Special stations for executing these movements are introduced at appropriate points. Again, some machining operations take longer than others, and it would obviously be wasteful if the majority of stations had to stand idle for half of each cycle, waiting for one or two slow-coaches to finish. This difficulty is overcome by what is known as "double indexing"[1]—by arranging for two successive stations to do the same job, each piece of work spending twice the normal time in one or other of these.

Clearly, if such a machine is to work smoothly, there must be elaborate arrangements for coordinating its various parts. If the transfer mechanism were to operate before a drill had fully withdrawn from the hole it was working in, the result would be a broken drill, and a serious hold-up for repairs. To avoid this sort of trouble, the machine makes use of feedback, although only in a rather elementary way. Each station is electrically connected to a main control panel. Only when every station has reported "Job completed and tools fully withdrawn" (and, of course, the two operators have also made their push-button reports that their jobs are done) can transfer take place. If a drill gets

1. To "index," in engineer's parlance, is to move a piece of work or a movable tool from one working position to another.

stuck in its hole, the machine stops completely, and the operators—or if necessary, the maintenance staff—must take appropriate action.

It is often necessary for the machine to check up on its actions in more detail than this. Frequently a hole that has been drilled at one station has to be tapped (have a screw thread cut in it) at a later one. If the hole were not drilled deep enough or if some swarf were left in it, then the tapping drill would be broken. To guard against this, a station in between is equipped with a probing device. It pushes into the hole a hollow rod, down which air is blown. The air blows out the swarf, while the probe must travel to the full depth before it can signal "All clear" to allow the next cycle to start.

Like any machine tool, the transfer machine is, of course, set in advance to turn out its work to specified sizes. But, again as in any other case, its tools wear and its working parts may get out of alignment, so that the work has to be inspected to insure that it keeps up to standard. This could be done by a human inspector, just as in preautomation days. But in all but the most elementary cases it pays to have the machine do its own inspection. A special inspection station at the end of the line applies gauges to all the important dimensions and, if they are outside the permitted tolerances, stops the machine and calls for attention. (It is possible to arrange for the inspection station in some cases to feed back instructions that cause adjustments in the machining stations, but, so far as I know, this is not normally done at the elementary transfer machine level.)

Let us now make a more detailed examination of the transfer machine for working Austin A40 cylinder blocks. This machine does not by any means do all the machining operations on the blocks—a great deal has to be done by other machines, mostly of old standard types—both before and after it in the line. Nevertheless, it does a very impressive list of jobs:

> Drilling all holes in the front and rear faces (except the oil gallery holes);

Core drilling, semifinishing and finish boring of the camshaft and crankshaft bores;
Facing, opening out, and reaming the oil pump bores;
Milling the cam clearance round the oil pump bosses; and
Facing the center bearing crank thrust face.

It used to require thirteen separate machines to do those jobs, and thirteen separate operators to attend them. Now the work is done by one machine, with two operators.

Transfer machines of the simple type so far described can be used for jobs that involve such operations as drilling, reaming, boring, countersinking, tapping, milling, grooving, and (with rather more difficulty) broaching—that is, operations in which the work remains stationary and only the tool moves. When turning or grinding is involved, the transfer problem becomes much more complicated.

But even with the restriction to "work-stationary" operations, the transfer machine has a wide field of application in, for example, the production of car engines. It can deal with most of the machining needed on cylinder blocks, cylinder heads, gearbox casings, exhaust manifolds and the like, as well as a number of smaller components.

In the U.S.S.R. a 16-station transfer machine does 504 operations on cylinder blocks for ZIS-150 lorries, turning out 30 blocks an hour, replacing 56 standard machines and reducing the floor space needed by 60 percent. Further detail is given by A. Zvorykin in *Engineering Progress in the U.S.S.R.* (Moscow, 1955). In the summer of 1955, I personally saw working at the Stalin Automobile Works in Moscow another cylinder-block machine of about similar capacity, and one machining gear-boxes, which replaces 15 standard machines. The general quality of Soviet transfer machines appears to be comparable to our own, though they seem to be rather behind us in the number so far in use (a different picture will emerge when we come to consider more complex types of automation).

A single transfer machine will not usually do all the operations required to convert a casting into a finished component. So the natural development is to link a number of machines for the purpose.

Transfer machines are connected together end to end, with automatic conveyors in between, to give completely automatic lines as much as a quarter of a mile long. There may be one laborer at each end, for loading and unloading, but otherwise there is no direct labor at all—only maintenance and supervisory workers.

Ford led the way in this development. The line for machining six-cylinder blocks at their engine plant at Cleveland, Ohio, is 1,545 feet long. On a series of 42 linked transfer machines it does completely automatically all the 530 operations and the many inspections which transform the rough casting into a finished block ready for assembly. Turning out 5,000 blocks a day, it needs only 41 workers, against the 117 used by earlier Ford methods, which were already very advanced.

The processes that we have dealt with so far involve only those types of operation in which, during machining, the work remains stationary. The transfer operation is therefore little more complicated than that of pushing a toy truck along a model railway and stopping it precisely at certain points. When turning operations are involved the problem of automatic transfer becomes much more complex. The manual operation would run something like this: lift the workpiece from the conveyor, place it in the chuck of the lathe (or between the centers of a center lathe), clamp it there, release the grip of the fingers on the piece, take the hand to a safe distance and start the machine; and at the end of the turning process, grip the piece between the fingers, release the chuck, remove the piece from the chuck and place it on the conveyor. To do the equivalent of these movements mechanically is obviously much more difficult than the simple transfer operations so far considered. Much the same applies to cylindrical grinding, while centerless grinding presents just as complicated, if rather different, transfer problems. Another problem arises with centerless grinding: the tool movements cannot be *prearranged* (as they can, for example, in turning) to reduce the piece to the required size; and so the grinder must be provided with a device which continually measures the diameter of the work and automatically stops the machine at the right moment.

It is not surprising, then, that (setting aside the degree of automatic operation used in the multispindle automatic lathe and its variants) the automation of processes involving turning and grinding should be a much more recent development than the simple transfer machine for cylinder blocks and the like.

The Pontiac piston line, which is completely automatic in all its machining and transfer operations and inspections, starts with rough-turned castings and ultimately delivers finished pistons. Omitting a few technical details, the operations performed are as follows:

> Core-drilling and other operations on the pin holes;
> Turning the head, grooves, lands, and skirt;
> Drilling smoke and oil holes;
> Inspection of skirt diameters and sorting into high- and low-compression types;
> Weight adjustment (the machine weighs the piston, mills the appropriate amount from a protuberance inside, and checks correctness of final weight).
> Inspection of the smoke and oil holes;
> Grinding the skirt;
> Tinplating (the first example we have met in this survey of a chemical operation);
> Finishing of pin holes; and
> Final automatic inspection (in a temperature-controlled room), together with sorting by pin hole size and skirt diameter.

The output is about 200 an hour. Automatic assembly is much less advanced.

Some hint of how future developments may go can be gleaned from the radio industry. Looking at the complicated mass of connections inside a standard radio set, one would be inclined to guess that it would be hopeless to try automatic assembly. But as soon as one works free of the idea of merely doing the assembly of an existing set automatically, and decides instead to redesign the set from scratch with automatic manufacture in mind, then the problem solves itself as if by magic. The basis of the solution is what is called the "printed circuit." One of the several variants of this technique starts with a molded bakelite panel on which the circuit is represented by a collection of grooves. After shot-blasting, a layer of metal is sprayed over the whole surface. Then the face of the panel is milled until metal is left only where the grooves originally were. Coils and condensers, as well as connecting links, can be automatically produced in this way. In the most recent developments, the panels then pass to a machine which automatically inserts most of the remaining components—though so far valves have proved too delicate to be treated in this way. Finally the soldering of the connections is done by automatically dipping the back of the panel in a solder bath. It has been estimated that automatic machines in existence today will ultimately supply half the small table sets used in the U.S.A. Similar procedures are also being used to assemble television sets.

Using this method, an American concern has set up a radio assembly in which two workers turn out a thousand sets a day—an output that would formerly have needed two hundred. So here we have a hundred-to-one increase in the productivity of labor—by far the largest we have met yet. The suggestion naturally arises that the problem of automatic assembly in general may be solved, not by automating existing assemblies, but by fundamentally redesigning the product, and that perhaps this can lead to comparable labor saving on other products.

An alternative method of building radio sets that also lends itself to automatic working is the so-called "modular design." In essence this consists in building up the circuits, as a child builds with toy bricks, out of ceramic wafers, each of which has an element of the circuit worked into it. This offers much less in the way of labor saving—perhaps only 50 percent—but it is much more suitable for producing a large variety of different models in small numbers.

The forms of automation described in previous pages apply only to large-scale serial production. This can be used for the production of prototypes or for "a-few-off" jobs. The other forms took over the work of unskilled or semi-skilled repetition workers. This can be used to replace highly skilled craftsmen.

Let us consider the case of a man milling a prototype—perhaps a complicated cam—from the blueprints. It is highly skilled work to read the information contained in the prints and translate it into terms of movements of the controls. It is usually also very slow work—with one job often taking several weeks. But when we examine what goes on, we discover that the machine is in use for only a very small part of the time. Most of the time is "thinking time"—time spent by the craftsman in poring over the blueprint and deciding just how to translate blueprint language into machine movements, and checking his decisions carefully since one mistake can undo the work of many days.

His work can be taken over by what is called a "program-controlled machine." Part of his skills have been built into the machine. But the chief novelty is that it has been equipped wtih sufficient "intelligence" to enable it to read a set of instructions that are presented to it—usually in the form of a list of measurements—and translate these into action.

The instructions are presented in the form of patterns of holes punched in a paper tape. The punching is done on a machine rather like a teletypewriter, and the skills needed by the girl who uses it are little more than those of an ordinary typist or a Hollerith punch operator—they can be learned in a few weeks.

This punched tape is now used to control the milling machine (or in some cases, the instructions are first transferred by an automatic device on to a magnetic tape—each system has its advantages). And now the machine runs continuously—there is no "thinking time." The control mechanism in the machine—its "brain," as it were—has been given the faculties necessary for translating tables of dimensions into motions of machine parts. For example, where an elaborate curve has to be milled, the instructions will only specify the positions of a number of points at intervals along it; the "brain" calculates how to join these up with a smooth curve and manipulates the motions of the machine accordingly. In more elaborate cases, the "brain" may make calculations about what cutting speeds are to be used for various types of cut and make the appropriate adjustments. This "brain," by the way, is a small-scale example of an electronic computer.

Feedback also plays a far more important role here than in any of the cases we have met previously. It is not possible to get sufficient accuracy by simple direct instructions to the moving parts. Instead, it is necessary to incorporate very refined measuring devices which report from moment to moment just where the various parts are; the brain calculates the difference between where they are and where they should be, and sends out further instructions to bring them into place. It would be impracticable to arrange the "brain work" of the control mechanism and the various feedback circuits in terms of electromechanical devices, and so at last with program-controlled machines electronics comes into its own.

For various technical reasons, program control has been applied first to milling machines, but it is equally applicable to all types of machine tools. *The American Machinist* (1st August 1955) described a program-controlled turret lathe. Formerly a machine of this type needed a skilled setter to prepare the job and a moderately skilled operator to run it. Now it runs itself, while

> the set up is planned by a methods engineer or "programmer" and the tape is punched accordingly by a clerk—both in the office. The tape is delivered to the machine with the shop order and operation sheet or tool scheme. An unskilled operator can mount the numbered, "plug-in," preset tools according to the chart and can place the tape in the machine director. As simply as this, the machine is ready to produce pieces.

An aircraft factory produces at the best only a hundred or

so of any given model; transfer machines are quite uneconomic for such short runs, but program-controlled machines could be used extensively. They would produce very great savings also when a mass-production factory is retooling for a new model. And there are possibilities that program-controlled machines may prove economical in small factories and shops where transfer machines would be quite out of the question, or in large factories which produce a great variety of products in comparatively small numbers.

Thus program-controlled machines promise great possibilities of a high degree of automation in almost any sphere, large or small, of engineering production. And increasing automation of long familiar machines points the same way. Take the well established automatic lathe, for example, so useful a tool for runs of a few thousand or more. Fed occasionally with a bar of metal, it will in its normal form automatically go through the motions of advancing the bar till the end is in the cutting position, bringing to bear on it a number of tools in succession, and finally cutting off the finished article and letting it drop into a hopper—after which the cycle is repeated. But tools wear down, and so the work has to be inspected frequently, and every now and then a setter must readjust the tools and eventually replace those too much worn by new ones. Now, however, the Sunstrand Machine Tool Company markets a lathe which measures each piece it turns off and automatically resets the tools to compensate for wear. And when the tools have worn down to a predetermined limit it automatically replaces them with sharp ones. Automatically loaded and unloaded in addition, these machines can run from five to eight hours without attention (except for an occasional check to make sure that parts are being delivered to the loading mechanism). Clearly any shop or factory that can use an automatic lathe at all, can use one of these with very great savings in the wages of both operators and tool-setters. Remembering that an automatic lathe can be changed quite quickly from one line of production to another, it is obvious that this type of machine offers great possibilities for many small and medium-sized shops.

So far as engineering production is concerned, the car industry has in fact led the way in automation—because it is by far the largest mass-production branch of that industry. But other branches of engineering are introducing automation too—though more slowly.

An American Ordnance Factory for making 155-mm. shell casings must have been easily the most advanced example of automation at the time of its completion towards the end of the last war. Starting with 24-foot bars of steel stock, it cuts the billets, forges the casings, and does all the machining operations on them. The work is untouched by hand, but many of the operations are manually rather than automatically controlled. The factory employs about 140 workers, including engineers and maintenance staff. A Westinghouse plant in Ohio uses a high degree of automation, including some automatic assembly, in its 2,000 odd machines and 27 miles of conveyors to produce five refrigerators a minute. The U.S.S.R. has an automatic line for making tractor ploughshares, which uses 15 workers instead of a former 150. It starts with sheet metal, cuts and rolls it, grinds the surfaces, subjects the shares to heat treatment, cleans and oils them, gives them an anti-rust coating, packs them and loads them into trucks. They have also applied transfer machines to making water taps and to drilling the holes in sewing machine bodies and inserting the various pins and bolts.

In the metallurgical field, highly automatic blast furnaces and rolling mills have been used in the best plants for many years, but recently many other processes have been made automatic. An American automatic iron-casting machine uses six men instead of 40 in turning out 3,600 two-and-a-half pound castings an hour. It can be changed from one job to another in half a minute. On a capital investment of 600,000 dollars, it saves 300,000 dollars a year in wages. Machines are now available, small enough and cheap enough to be used in the average foundry shop, which will make cores and molds in half a minute that used to take a skilled man a quarter of an hour. The continuous casting process, in which the molten metal pours into the top of a long vertical tube and cools on the way down so that a continuous billet emerges at the bottom, was developed for brass in the early 'forties and applied to steel in 1948. Besides being substantially automatic, it gives very large savings in cost by eliminating several stages of the orthodox process and producing a semifinished product in one operation.

Recent developments in sintering (a sort of casting process in which finely powdered metals, pressed and heated in a mold, are partially fused and coagulated to form the finished article) are making a strong bid to eliminate machining operations altogether from the production of small components. For sintering can now produce these to accuracies of a thousandth of an inch and with tensile strengths only slightly less than those given by standard methods. This is not necessarily an automated process in itself (though it could easily be automated), but it will have the same effect as very thorough automation of the machine line —for the metallurgical process uses far less labor and far less elaborate machinery than metal-cutting.

Turning to the electrical industry, we may note an American machine which makes electrical resistors, calibrates, classifies, and packages them with virtually no operator attention at a rate of 3,000 an hour—formerly the output of 82 people. Another machine making thermostats occupies ten square feet of floor space in place of a former 6,000 and enables one man to turn out as many instruments as 22 used to do. On the electronics side, we have already seen the rapid progress of automatic assembly of radio and television sets. And it only remains to add that the manufacture of components like valves and cathode ray tubes has also been made substantially automatic.

Automation is, in fact, cropping up in all sorts of industries. On one floor of a Columbia Record Company factory, 250 men work old presses producing gramophone records; on the floor below, with automatic presses, four men turn out eight times as many records—a productivity increase of five hundred times! A British chipboard factory is virtually automatic in operation. The raw material, which may be large logs of wood or small pieces, is ground and then the chips are carried pneumatically to a screening machine and thence to a drier in which their final moisture content is automatically controlled. The next conveyor acts as a weighing machine, and as they leave this the chips are sprayed with a mixture of resin and hardener in automatically controlled proportions. The "carpet" of sprayed wood passes continuously through a press where it is heated and finally formed into the chipboard. In a new factory of Micafine Ltd., at Derby, five men per shift supervise machinery which turns out over £300,000's worth of mica dust per year.

One might think that automation is unlikely to make much progress in industries like transport and building. Yet automatic marshaling yards are already established on several U.S. railways. In the one at Conway, Pennsylvania, every wagon on a train is sent to the right siding by switches controlled by an electronic "brain" fed with information on a punched tape. As each wagon or group of wagons rolls down from the hump, devices incorporated in the tracks weigh it and also determine how much it is being slowed down by friction. Other apparatus notes how many trucks there are already in the siding and calculates how far the new group has to run. A sort of radar device measures their speed as they approach. The combination of all these pieces of information controls retarding devices on the track which act to bring the wagons to a stop at just the right place.

In the building industry there is obviously great scope for automation in the manufacture of prefabricated components—the chipboard factory we have just discussed is a case in point. Concrete mixing has been automated in both the U.S.A. and the U.S.S.R. The American plant can provide 1,500 different mixtures as required. The operator selects from a file a punch card representing the required formula. When he has placed it in the control mechanism, the rest is automatic. Specified quantities of the materials are delivered to a mixer, mixed, and loaded into trucks for dispatch. The machine produces 200 cubic yards of concrete per hour. The Russian plant, for use in connection with hydroelectric and canal construction schemes, is made in sections so that it can be easily transported. It uses no manual labor, and has a staff of 17, whose working life is spent at control panels pressing buttons. The process is substantially automatic from the unloading of the raw materials from trucks to the placing of the finished concrete, and the production rate is 4,000 cubic meters a day.

It is all very well to automate the production of building materials and prefabricated parts. But could one hope to do the actual building automatically? Fantastic as it may seem, this has already been achieved on an experimental basis in Warsaw. An electrically driven bricklaying machine, operated by one man who needs no bricklaying skill, spreads the mortar and places the bricks, leaving openings as desired for doors and windows, and building walls of any specified height and thickness.

In the food-processing industries, automation is used not only to save labor and costs, but also to improve standards of cleanliness. In many canneries, sheet metal, paper, cardboard, and the produce to be canned, fed into their respective parts of the mechanism, are automatically transformed into cartons packed with sealed and labeled tins of sterilized food. Six largely automatic bakeries (the first of which was built in 1931) now provide almost the whole of the bread used by Moscow's population of eight million plus.

These more or less random examples will be sufficient to show that, despite the automobile industry's present lead, automation has a very wide field of application. There are few major industries that will not be affected by it, and those that have lagged so far may be expected to catch up in the future.

Conditions affecting automation in the chemical industries are very different, both technically and economically, from those in engineering. On the technical side, the problems of automatic transfer are usually very much simpler. Most chemical operations are done on fluids (liquids or gases), or on solids that are made to behave like fluids by powdering them and carrying the powder about in streams of liquid or gas. Arrangements of pumps and valves, which are quite simple in essence, will do all that is required in the way of transporting such materials from one part of the plant to another.

As against that, problems of process control are usually more complex in chemical production. The control involved in machining some article from metal usually amounts to little more than ensuring that at each station some dimension is reduced to a given size by the time the operation is finished. But to ensure that the right end-product comes out of a reactor vessel in a chemical plant, a whole series of separate factors must be simultaneously controlled within fine limits—the temperature and pressure in the vessel, the rates at which several reagents are fed in, the level of the liquid lying in the bottom, the rate at which the product is drawn off (or the several products if the operation is a distillation in which quite different substances condense in different parts of the tower), and so on. Furthermore, these various factors will usually interfere with one another —an increase in the input of some cold material will tend to lower the temperature, and so more heat must be supplied.

Because of these differences, the operating staff in a modern chemical factory (say an oil refinery) perform very different functions from those in an engineering factory. Since "transfer" is done by pumps and valves, the operators are not to be found near the work itself. Instead, they are concentrated in control rooms, surrounded by dozens, perhaps hundreds, of instrument dials, switches, and control wheels. Their job is to watch the movements of the needles on the dials and to take appropriate action from time to time in order to keep each needle pointing where it should. If temperature rises, then a knob must be turned to reduce the flow of the steam that does the heating. If a liquid level drops, the rates of input and withdrawal must be adjusted accordingly.

Even the most advanced chemical factories still fall a long way short of completely automatic operation. Only individual stages—like distilling or cracking—out of the whole complex of processes are made substantially automatic within themselves. But each stage feeds the next, and what is happening at each stage affects operating conditions in the next. At present the adjustments that this requires have to be made by operators. A completely automatic plant, in which all stages of manufacture were subjected to one integrated control would require, not only a control of the "electronic brain" type (which might perhaps be made available even now), but also a great deal more knowledge than we now have of exactly how the varying conditions affect the output of each chemical process, and very great advances in measuring instruments.

Obviously what the instruments should measure is the quality of the product—say, at the final stage, the octane value of petrol produced. Instruments that will do this are not yet available in most cases. The automatic controls in use at present simply ensure that temperatures, pressures, rates of flow and the like are kept at levels which are expected to give the required quality. Samples of the products are taken at frequent intervals and sent to the analytical laboratories for checking. If the quality proves to be wrong, the operating conditions will have to be altered, and instructions go to the operators to make the appropriate adjustments—after which the automatic controls operate at new levels. All this takes time—during which the plant has been producing an inferior product. Obviously it would be advantageous to automate this quality control, so that the plant would be adjusted automatically and nearly instantaneously to keep the product at the right quality level. That will presumably come some day—an automatic chemical laboratory might feed back instructions to the plant controls; infra-red spectrometers can perform certain analyses continuously on moving streams of fluid.

The car industry employs three hundred thousand or so manual workers in this country. Against that we have about 2½ million clerks. So the automation of office work, which has been going on at an increasing pace for the last few years, might well have far greater effects—for good or ill—than any of the machinery we have so far surveyed.

A great deal of office work can be reduced to a simple routine. That is to say, it would be possible to prepare a set of instructions for the day—or the week or the month—such that if the clerks merely follow the instructions accurately, without ever taking decisions on their own, the work of the office will be done. The instructions may well have to include alternative courses of action, but these, too, can be reduced to instructions to be mechanically obeyed—"When you get to this point, if there is a credit balance, do this; if a debit balance, do that." Work of that type ought to be easy enough to automate.

In some respects office work is easier to automate than anything we have dealt with so far. When production of real articles is to be automated, we have to face the fact that real materials have awkward properties—all those metal-cutting machines, for example, produce swarf and chips, and the clearing away of this waste so that it does not foul the machines is one of the biggest headaches in engineering automation. Again, no matter how carefully they may have been manufactured, real materials vary in properties from sample to sample, and the machines must be designed to allow for these variations.

But the figures with which the clerk works are abstract symbols, free from all the cussedness of real materials. True, they have to be represented by something material before they can be manipulated—by conventional marks on paper, for instance, in the established practice of centuries. But we can choose what sort of material representation we will have, and we can choose it in such a way that it will be easy to manipulate automatically. One useful method is to represent numbers by the positions of gear wheels—a ten-toothed wheel advanced seven teeth beyond its zero position represents the digit seven. Out of combinations of such wheels there emerges the ordinary desk calculating machine, which mechanizes some part of the clerk's work—mechanizes only, not automates, since the machine must be controlled in detail by its operator.

But gear wheels have far too many awkward material properties—considerable power needed to drive them, backlash which provides problems growing more and more serious as the chains of gearing get longer, and so on. And gear wheels move slowly. A much better way would be to represent numbers by pulses of electricity. Modern electronic devices can manipulate such pulses with great speed, great accuracy and great reliability. And it is along these lines that the "electronic computer" (popularly known as the "electronic brain") has emerged.

This is not the place for a detailed description of how an electronic computer works. It will be sufficient to give the merest outline. As already indicated, the numbers are represented by electrical pulses circulating in various circuits. There is a "store" or "memory" in which the machine can keep a record of numbers it will need to use in the future. There are a number of arithmetical units which perform the ordinary arithmetical operations. An adding unit, for example, will be a combination of valves such that when it receives two groups of pulses representing number x and y, it sends out in response another group of pulses representing $x+y$.

And lastly, there is some form of master control, by which the various parts of the machine are told at every instant just what they must do next: "Memory! Send this number and that number to the adding unit. Adding unit! Feed back the result to this (specified) place in the memory." And so on. It might seem at first sight that nothing important has been gained if the machine has to be instructed about every step it can take. But the point is that any lengthy calculation can always be reduced to a routine which is to be repeated over and over again. The machine is fed with a certain set of numbers to start on. It is told to perform certain calculations on these and store the results in the memory. It has then to do the same routine of calculations again on a different set of numbers, some or all of which will be numbers that it produced in the first routine. And so on. Occasionally there may be need for varying the routine—just as the clerk must take different courses of action according to whether a balance is credit or debit. But instructions that allow for that can easily be arranged. Every set of instructions, in fact, must contain at least one such provision—an order to stop when the calculation is finished.

It must not, like the broom in the *Sorcerer's Apprentice,* go on forever when the job has been done. The criterion that decides when the job is done may take various forms—"Stop after 783 repetitions of the routine," or "Stop when a certain cumulative total has passed a given figure," or "Stop when a number that is being calculated by successive approximations has reached a given degree of accuracy." So the routine instructions incorporate an order of the type: "At the end of every cycle, examine a certain number; if it is (for example) positive or zero, repeat the routine; if it is negative, stop."

Usually the numbers that the machine is to use and the instructions it is to follow are fed into it in the form of patterns of holes punched in a tape (or cards). Having read its instructions the machine sets to work, carries out the complete calculation automatically and eventually puts out the results. The output may be in the form of figures printed by an automatic typewriter, or in the form of yet more punched tapes or cards to be used in future calculations, or (quite commonly) both.

The idea of a completely automatic machine to do all this is quite an old one. It came almost fully fledged from the brain of Charles Babbage, one of the nineteenth century's most frustrated geniuses, in 1833. But at that time, only mechanical devices like gear wheels, ratchets, and levers were available for putting it into practice. And they are far too clumsy for the job, so that Babbage's attempt ended in complete failure.

It was the development of electronics that made it possible to turn Babbage's ideas into working computers.

These machines work at enormous speeds—several tens of thousands of additions, subtractions, multiplications, and divisions per second are now the usual thing. That does not merely mean that they can do calculations faster than human computers; it means in practice that they can do calculations that would not be possible at all without their aid. For there are many scientific problems in which the theoretician can write out a set of equations and say: "The solution to these is the answer you want." The only trouble is that to solve the equations without the aid of the electronic computer would often take generations, or even centuries. We cannot wait for that—and so till a few years ago these problems were in practice insoluble. But the electronic computer nowadays runs off a solution in a few hours, and in that way is enormously accelerating the progress of many branches of science.

The electronic computer proves in practice to be not only very much faster, but also considerably more reliable than its human equivalent. It has its breakdowns, of course, and an efficient maintenance service is essential, but by and large it makes far fewer mistakes than do human computers. And it is a great deal cheaper—even with the first of these computers ever produced, when interest, depreciation, running costs, staff salaries and all the rest were allowed for the cost of doing a calculation worked out at something like an eighth of what would be required using human computers with desk calculating machines. And enormous improvements have been made since then.

Apart from military applications—in ballistics, for example—electronic computers have been widely applied to the extremely complex calculations that occur in such fields as X-ray crystallography, atomic physics, optical design, calculations of stresses in engineering structures, aerodynamic design of aircraft, weather forecasting and many types of statistical work.

All these jobs are essentially mathematical ones, and it is true that up till recently these machines have been used chiefly to carry out calculations—hence the name "computers," from a Latin word meaning to reckon or calculate. But that is, in a sense, a historical accident, and these are really machines for handling large quantities of any sort of information that can be precisely stated—for sorting it out in any way that may be desired, for deducing its logical consequences, and for presenting the information in any way that may be found convenient. They can be made, for example, to play chess—rather poorly, it is true, for though their logical analysis of the various possibilities in the next few moves is unimpeachable, they lack any ability to comprehend a strategic situation as a whole in the way that the experienced player can do. At the present time, several teams are developing them as translating machines; their job here will be to accept the information conveyed by the words of one language, refer to a dictionary incorporated in the "memory" to find the corresponding words in the second language, work in the other pieces of information that are conveyed by such things as inflections and word order, and finally turn out a readable translation of the original. To emphasize the fact that these machines possess abilities much wider than that of mere rapid calculation, a recent tendency has been to refer to them as "data processing machines" or "data processing equipment," though the term "computer" is holding its own, even for cases where actual computation plays a minor role.

From all this general description, it will easily be seen that electronic computers should be able to take over many of the tasks that are normally done by teams of clerks in large offices.

The machines are simply doing more quickly and more cheaply jobs of types that armies of clerks have been doing more clumsily for many years. But the speed of the electronic machines is already making it practicable to undertake new types of work that have previously been quite impossible. This particularly applies to tasks of assembling and sorting information on which to make managerial decisions about production policy.

A computer has been installed at the Fawley refinery, the main task of which will be to facilitate the selection of a production plan that will meet as economically as possible the changing market requirements from existing stocks of crude and partially processed oil. The data—the orders and the stock figures—are available, of course; but human computers could not do the sorting and calculation quickly enough for it to be of any use. On the other hand, the electronic computer will produce not just one but several alternative production plans in a matter of hours—leaving it to the management to make the final choice. Other production control problems that have been solved by computers include such things as: minimizing transport costs of a firm that has eight large factories and 56 warehouses, with a choice as to which factory shall supply which warehouses with which products; planning the most economical use of the limited resources of a farm, given the market conditions; and discovering which of several government contracts offered to an aircraft firm would pay best in terms of the known disposition of the firm's resources.

If computers can assemble and analyze the information needed to make production-control decisions, might they possibly go further—might they receive their information automatically from its sources, draw their conclusions from it and themselves send out to the various machines the instructions that must be followed? Certainly there are instances of computers—much simplified computers—directly controlling machine behavior; the program-controlled machines and the concrete mixing plant. But these are cases in which the program is supplied from outside and the computer merely interprets it in terms of machine operations. Something much more advanced has been achieved in electric power production, which is, of course, one of the simplest of all processes to automate. For many years now, the operation of the individual powerhouse has been largely automatic—once the engineer in charge has prescribed the conditions, automatic controls take over. But a major problem is to prescribe the regime that will give the most economical working of all the stations in a large grid under the varying load conditions. Electronic computers can now solve that problem. Given the pattern of demand at any moment (supplied automatically by meters) they can calculate how much power each station should be producing in order to give the most economical result, and automatically transmit the appropriate instructions to each of them. Already a computer controls in this way the operation of 35 generators in the nine plants of the Ohio Edison Company scattered over 9,000 square miles of country, and the U.S.S.R. has similar but more ambitious plans for the automatic control of the Moscow grid, which includes stations 600 miles away in Kuibyshev and Stalingrad.

Electric power generating, of course, presents these control problems in their simplest form, and it may be a good many years before we see computers running engineering factories or chemical plants. But looking ahead, it is possible to visualize computers not merely operating plants more efficiently and cheaply than human managements, but even making it possible to work processes that could not be attempted now. Some theoretically possible chemical processes, for example, are beyond our reach at present, because they tend to get out of control and "run away." No human operator could think or act quickly enough to keep them stable. But an electronic computer, fed by signals from measuring instruments throughout the plant, could correlate the relevant pieces of information and send back appropriate control signals in a thousandth of a second, and in this way make such processes practicable.

It goes without saying that they can never do any really creative thinking, never take major decisions. They can examine data presented to them and use it to "decide" on one course of action rather than another—but only on the basis of instructions that human beings have prepared for them. Their "decisions" are completely automatic, wholly unintelligent. Unless some human brain has foreseen and prepared instructions for every possible eventuality, the computer will be lost. It has no faculty for dealing with the completely novel. That is reserved for humanity.

But what the computer can do for us—and do far, far better than we can do ourselves—is to perform low-grade brainwork in a high-speed way.

It is often said that automation is not just the substitution of automatic machinery for human labor and control, but is rather a "new philosophy of production"—meaning that automation is not to be achieved merely by accepting existing processes and making them automatic, but should rather recreate the whole manufacturing process on an entirely new basis.

I forget (and I am not going to search the reference books to rediscover) how many years elapsed after the invention of the incandescent gas mantle before it struck somebody that it would be better to have jet and mantle point downwards, instead of in the skyward direction that had been inherited from the days of the naked flame. The early motorcars were truly "horseless carriages," very little different from the horse-drawn vehicles they were destined to replace, only slowly were they modified to a design really apt for self-propulsion—and indeed, many people doubt whether, in the matter of external contour, they have reached that design even now. It has been the same whenever new manufacturing methods have come along.

And the same thing seems to be happening—must inevitably happen—in the early days of automation. We used to make cylinder blocks on lines of machines with manual transfer between them. And so we take our first steps in automation by pushing the machines closer together and mechanizing the transfer. We do not stop to consider whether a different design of block, or a different design of the whole engine, might enable us to use some far more efficient automatic method. Yet the present form of the block has been determined just as much by the methods formerly available for making it as by the function it has to fulfill. And the same applies to the design of the whole car—or of any other mass-produced article. This design has been profoundly influenced by the whole structure of the method of interchangeable manufacture as it has been practiced for the last thirty years or so. The machines which are automated are, in their essentials, the old machines with a few frills added—program control is applied to standard milling machines; an automatic lathe is equipped with the addition of an automatic tool-changer; transfer devices are added to only slightly modified standard machines. Automatic machines designed anew from scratch and free from the influence of now irrelevant traditions have hardly begun to appear.

There are plenty of designs and processes excellently adapted to the production methods of yesterday and today, but completely unsuitable for automation. Could one conceive of any practicable and economic method of automating the upholstering of car seats (or drawing room furniture) with their present designs and materials? Or would one dream of trying to weave the homely shopping basket automatically from its reeds? Obviously not.

Eventually, automation will not be a matter of merely substituting automatic devices wherever human labor is used at present. It will be a method of manufacture as different from the established methods of today as the latter are from the handicraft work of two hundred years ago. And presumably automatic manufacture will in due course replace present manufacturing methods as completely as these have replaced handicraft production. But does "in due course" mean in ten years, or in thirty, or in a hundred? Nobody can yet say.

The last paragraph raises the question of the automatic factory. Can we expect a time to come when most of the material things of life will be produced by processes that are substantially automatic from raw material to finished article? Before we try to answer that question as a whole, let us look at the weak link in the chain of automation as it is today. That main weakness is undoubtedly in assembly.

The car—as usual—is the most convenient subject to discuss. At present its 5,000 or more parts are assembled by many hundreds of workers ranged along the conveyors of a final assembly line and of the many subsidiary assemblies that feed it. If we are to visualize automatic car factories, then we must first see if that assembly can be done automatically.

So, in view of what has been said above, our thoughts turn first to the question of redesign. The consumer certainly does not want a car of several thousand parts. A few dozen will satisfy him—wheels that revolve are the first essential, and after that come the various controls (some of which he hopes to see eliminated by things like automatic transmission), doors and the handles that open and shut them, sliding windows and the knobs that move them, and a few other incidentals. As far as the user is concerned, the rest of the vehicle can be made all in one piece. Of course, the automobile engineer insists on more than that, for the car has to work.

But is a total of 5,000 or more parts essential? Of course not. Surely it would be possible to construct the same body out of a few large panels, or even in one piece?

The electric wiring might be automatically molded inside the body work, which would do away with a great deal of electrical assembly work, as well as with the need for the ordinary insulating coating on the wires.

I do not know how many parts make up the engine—several hundred, I should imagine. Many of them will be essential to the working of the engine, but presumably many others are merely manufacturing conveniences of the present, which become manufacturing inconveniences from the point of view of automatic assembly and which could be eliminated by redesigning.

Then there are all the accessories. Each would have to be considered on its merits to see if its construction could be simplified or if by some means (other than the obvious one of doing without conveniences) it could be eliminated. Take the windscreen wiper, which must, as it stands, involve some dozens of assembly operations. Glass coated with certain compounds of the silicone family has the property that any water on it forms into large separate drops. Could one redesign the shape of the windscreen (and if necessary the bonnet) on aerodynamic lines, so that when the car is in motion these drops are swiftly blown aside to leave a clear view? Or, if that does not work well enough (for example, at low speeds), could one find a coating with the opposite property, a coating which would make raindrops spread out into a continuous even film over the whole windscreen? That would solve the problem, for it is all that the wipers do now.

In detail these suggestions may all be nonsense. They have simply been put in to suggest that if engineers consider car design carefully, resolutely abandoning traditions that arise from the needs of present manufacturing methods, and grasping at the same time at every new possibility that modern science and invention can provide, then they may end up with a car consisting of two or three hundred parts rather than the several thousand of the present time. And then the prospect of automatic assembly becomes a great deal rosier.

Given time—and given economic conditions that provide the necessary incentives—all these problems could be solved. There remains one major difficulty. Even when assembly is done by manual labor alongside conveyors, the problem of bringing components to the correct places as and when required is a serious one. The hundreds of minor fittings must be coordinated so that final assembly can run smoothly. The problem was not so serious in the days when Ford could say that his customers could have the T-model in any color they liked provided it was black. All T-model cars were absolutely identical (except that one might choose between a roadster and a coupé body), so that it was only necessary to ensure a continuous flow of all components to the appropriate stations. But nowadays a single line will be producing many versions of the same basic model—with variations of body type and color, different upholstery schemes, left- and right-hand drive, perhaps engines of different capacities and choice of synchromesh, preselector and automatic gear-box, as well as various optional fittings from overdrives to radios and heaters. Almost any combination of these may occur, successive cars on the line involving quite different assemblages. If twenty components could be varied, with a choice of two possibilities for each, the total number of different combinations would be rather over a million, and it might take a year or two for any given combination to crop up a second time. In practice things are not quite so difficult as that. But they have none of the simplicity of the T-model days. It is no longer enough to ensure a steady flow of each component to the place where it will be wanted. Instead, the material-handling side of the factory organization must arrange that each variation of each component (coming from some subsidiary line or from store) meets the chassis for which it is intended at the appropriate place on the line and dead on time.

To cope with this sort of thing, Austins have embarked on an ambitious scheme for controlling the movements of the various parts by a Hollerith punchcard system. Such a control system is well within the reach of an electronic computer.

And now we really are within sight of the automatic factory. Whatever the works takes in from outside would be tipped into hoppers or picked up by automatic handling devices and loaded on to conveyors. The aim would be to do as much as possible within the walls—to receive metal ingots, for example. Many of the metallurgical processes and all the machining could be done automatically, while jobs like upholstery that obstinately defy automation will have to be replaced by more suitable alternatives. And so the work would flow, without human intervention, from machine to machine. Storage hoppers would intervene here and there to cushion any stoppages or to facilitate the problems of general coordination. From the machining lines the parts would flow to plating baths, paint-spraying chambers or other finishing shops, and thence (with or without further intervening stores) to the subassembly lines where they would be built up into engines, gear-boxes (if these still survive), back-axles, bodies, and the like. And these in turn would finally come together on the automatic final assembly line, at the end of which the first human hands that the growing car has encountered would press the starter button and grasp the wheel to drive it away.

All the many machines in such a factory would have to be kept carefully in step, and this would be the task of an electronic computer, or perhaps of several computers each controlling a group of related machines and the relevant subassembly, and all controlled in turn by the master computer. Every machine would be sending a stream of signals to the computer, reporting on its rate of output and its needs for raw materials. The computer would digest this mass of information and then send back appropriate instructions—this machine or department is to slow down, or that one speed up to bring them back into step; this group of multipurpose machines is to change from one line of production to another.

Then there would be a management grade to take major

policy decisions, for no automatic device can decide a policy that is intended to serve humanity. Apart from the possibility of having to deal with unforeseen emergencies—and they could never be entirely eliminated—management need not be concerned with anything less than top-level policy.

Somewhere in the background would be the really important people: the machine builders, the designers, and the research and development staff. The machine builders would never be idle, for there would be a steady trickle of demand for replacements or for new machinery to put into practice the improvements in either product or process that the research and development staff would think up from time to time. And even if the machinery should be built by automated processes, there must somewhere be the workers who will build the automatic machinery that will build the automatic machinery that will build the production machinery.

All this, I am sure, is not idle dreaming. My sketch may be wrong in almost every detail, but I am confident that production on something like those lines will some day be achieved.

> The twentieth century has seen the opening of a new era in construction, in which American engineering is a leader. Population and standards of living have increased in staggering proportions. Public and private transportation and communication have blossomed into activity hitherto unimagined. Practical problems involving power, materials, and methods have been attacked by a growing number of professional engineers. New alloy steels, glass, aluminum, and plastics have entered the market. Reinforced concrete and more recently prestressed concrete have been used in cheaper, sturdier, and sometimes startlingly unusual designs.
>
> Although its origins lie in the nineteenth century, the structure which perhaps best epitomizes this new development is the skyscraper. Sharply rising land values in New York and

Chicago, cheap-construction steel, and the invention of the hydraulic elevator were the elements out of which it was born. The stone walls of early multistoried buildings were under the handicap of having to support themselves, and were unconscionably thick at their bases. With the substitution of steel frameworks, walls were hung on internal structures which were sturdy, light, and easily assembled. Thus separated, the skin and the bones provided new opportunities for both architects and engineers. Some of the results are described in this article by a Professor of Architecture at Columbia University.

SKYSCRAPER: SKIN FOR ITS BONES

JAMES MARSTON FITCH

AS THE term is used nowadays in the United States, "curtain wall" describes the nonstructural sheath or skin which encloses our skeletal structures. In appearance, this new wall is quite unlike its predecessor, the load-bearing wall of history; and this novel appearance is not deceptive, for the new wall marks a big conceptual advance over the old one. The specialization of members involved in this separation of building skeleton and building skin corresponds to the specialization of tissue in biological evolution. And it has had equally powerful consequences, for the evolutionary step from the load-bearing to the nonload-bearing curtain wall has made possible a new order of performance in buildings.

Any architectural structure has two distinctly different tasks to perform. The first is simply that of carrying vertical and horizontal loads. The second is that of providing an enclosure which can regulate the flow of heat, cold, light, sound, air, and water between the building and its external environment. These two tasks create the age-old paradox of architecture because, generally speaking, they are mutually exclusive. That is to say, a material suitable for one function is not adaptable to the other. Steel and concrete are excellent load-bearing materials, but they are exceedingly poor thermal insulators. Mineral and glass wools make excellent thermal and acoustical insulators, but they have no structural value at all. Glass is transparent to light and heat and, for that very reason, is useless as an insulator. Wood is strong in tension and compression but vulnerable to fire and rot. There is, in short, no such thing as a universal building material.

The only way to escape this paradox is the way of nature —i.e., through structural specialization. But before the appearance of steel and reinforced concrete the only natural material available for specialized use was wood. Wood has been scarce in most of Europe for centuries; even in those areas where it was still plentiful, the hazards of rot and fire tended to limit its use to upper floors and roofs. Thus masonry remained the basic building material of western Europe. Since it was undifferentiated tissue—serving both for support and for enclosure—masonry could not perform either of its tasks very well.

American experience proved to be somewhat different. In the new world the English settlers found limitless supplies of virgin timber. They brought with them a familiarity with the relatively advanced, wood-framed house of the seventeenth century. These two factors combined to make wood the dominant building material on this side of the Atlantic. The English frame was subjected to steady refinement and produced, by the mid-eighteenth century, a highly specialized structural system: a light wooden skeleton sheathed in two skins—overlapping wood shingles or siding on the outside, plaster on wood lath internally. This familiarity with skeletal construction was to stand Americans in good stead when iron and steel became generally available in the last half of the nineteenth century.

Nevertheless, all the early metal skeletons appeared first in Europe. The English train sheds of the 1830's and 1840's, like the Library of Ste. Geneviève in Paris (1843), employed metal skeletons to achieve huge vaults of practically no mass and no weight. And this principle was shortly to be carried even further in two English structures of breathtaking scale and lightness—Decimus Burton's Palm House at Kew Gardens (1845) and Joseph Paxton's Crystal Palace in London (1851). Here the consequences of structural specialization, of the separation of building tissue into skeleton and skin, were clear for all to see. Although the vitreous skin did not yet protect the ferrous skeleton, the concept of the curtain wall had appeared in full flower.

Its origins were European, yet the subsequent development of the curtain wall was largely to be carried on by the Americans. There were many reasons for this. The first was, as we have seen, our familiarity with skeletal structures in wood. The second was our discovery, during the nineteenth century, of immense supplies of coal and iron ore. These made the metal skeleton as cheap, relatively speaking, as the wooden ones. But the third and greatest reason was undoubtedly the skyscraper. This characteristically American building type has dominated our skylines and our thinking since the 1880's, when the rapid growth of our cities made necessary—and the elevator made possible—this peculiar form of urban concentration.

As a matter of fact, the demand for multistory office buildings was so great, during those early days, that the first skyscrapers were not skeletal at all. Root's Monadnock Building (Chicago, 1891) climbed to a height of 16 stories with solid, load-bearing masonry walls. To reach this height, the walls had to be six feet thick at sidewalk level. But this was the absolute limit. Someone had to perfect a stable, self-supporting steel skeleton which could rise indefinitely, independent of enclosing walls. As a matter of fact, someone already had: William Le Baron Jenney built the Home Insurance Building in 1883—the world's first completely articulated, multistory steel skeleton clothed in a nonstructural skin.

Once this structural system had been invented, architectural attention could be focused on the further refinement of its parts. It must be confessed that during this period the architects' principal concern was with the skin: even here they saw the problem as more esthetic than functional—i.e., the emphasis was more on its *appearance* than on its *performance.* The principal walling materials then available—cellular terra cotta tile, brick, plate glass—produced walls which were weatherproof, fire resistant, and relatively lightweight. Such architects as Root (in his Reliance Building, 1895) and Sullivan (in his Schlesinger Building, 1899) used them wtih great distinction. But for the climates of North America such walls were far from ideal. A typical Chicago year presents temperature extremes unknown in Europe—from one hundred degrees Fahrenheit to twenty or thirty degrees below zero; a typical day can easily have extremes thirty, forty, or even fifty degrees apart. Solar radiation is intense in summer; heavy snows, high winds, and months of frost mark the winters. An American wall must be designed to meet both subtropical and sub-Arctic conditions. All this implies a high degree of specialization within the tissue of the wall itself (again the biological analogy!)—thermal insulation, *soleil brise,* waterproofing, vapor barriers, ventilation, daylighting, etc.

Another negative aspect of these early curtain walls was economic. While incomparably lighter and less massive than load-bearing masonry, they were still built up of relatively small units (brick and tile), whose assembly required a lot of hand labor in contrast to the skeleton itself, which was largely prefabricated. The next logical step, therefore, would have been to abandon masonry altogether and prefabricate the curtain wall in story-high, bay-wide panels. Only thus could the advantages of industrial production, already applied to the fabrication of the skeleton, be brought to bear upon the fabrication of the wall.

Yet half a century elapsed between the Schlesinger Building and the first prefabricated curtain wall. The reasons for this delay were two: the trade unions and the municipal building codes. The mason crafts, early and powerfully organized, were umbilically tied to the old techniques and naturally resisted technological change. The building codes, drafted during those same years, had a built-in prejudice in favor of masonry and ceramic walling materials. Their specifications were all written in terms of how a wall should be built, rather than in terms of how it should perform. It took decades of pressure to overcome these two obstacles, with the result that a nonmasonry curtain wall has become legal in most American cities only since the Second World War. Thus only in the last decade has it become possible to clothe the American skyscraper in skins of stainless steel, aluminum, porcelain enameled steel, and plastics.

Progress since the war, however, has been spectacular. The new metals, together with the new thermal and acoustical insulations, synthetic finishes, and insulating glasses, are yielding walls three inches thick (instead of 12 to 15 inches) which weigh 12 pounds per square foot (instead of 125 pounds). The panels are so large and light that one entire 23-story skyscraper in New York was sheathed in a single day!

In terms of its total area, glass has always played a dominant role in the curtain wall. Today's "crystal towers" employ only slightly more glass, perhaps no more glass, than some of Sullivan's buildings of sixty years ago. The glass itself has been importantly modified, however. The industrial production of rolled glass in large sheets was accomplished by the early 1880's, and the Chicago architects were quick to use it—in fact, big plates of fixed glass were originally called "Chicago windows." But ordinary glass, for all its remarkable properties, displays many deficiencies when used in the skyscraper. Its transparency to visible light and infrared radiation is both a blessing and a curse. In the bitter American winters, large areas of ordinary glass are uncomfortable, wasteful of heat, and cause serious condensation. In the torrid summers, on the other hand, they admit too much solar heat. To correct this the insulating glasses were developed—sandwiches of two sheets of glass separated by a hermetically sealed air space. They are 51 percent more efficient as thermal insulators than ordinary glass.

Another limitation of ordinary glass is its optical properties. Under most conditions, it is *too* transparent to visible light, offering no means of modulating its intensity or distribution. Early skyscraper architects soon discovered that, if the window was to be enlarged to cover the whole wall, then interior blinds or curtains or outside awnings became more important than ever. To meet this difficulty, glass manufacturers began to modify the light-transmitting qualities of the glass itself: the current tinted glasses reduce light transmission by 27 percent. When these glasses are used as the outer leaf of an insulating sandwich, a reasonably good control of heat and light is achieved. This sandwich was used in the first postwar skyscraper to have a completely nonmasonry curtain wall—Belluschi's Equitable Building in Portland, Oregon. It is this glass which gives so many recent skyscrapers their characteristic blue-green color.

This particular formula, far from being the final solution to the curtain wall, has proved to be merely the beginning. During the years when men were advancing the use of glass, many things were happening inside American buildings which were placing ever greater demands on the performance of the wall. New standards of human comfort and of industrial efficiency were demanding ever more precise control of the thermal, acoustical, and luminous environments. Since, as we have seen, the fluctuation in most American climates is so immense, the concept of the hermetically sealed wall appeared. In some factories, this concept went so far as to eliminate glass altogether—i.e., the skin was made as nearly opaque to all environmental factors as possible. The sealed-envelope concept has certain virtues for such industrial processes as pharmaceuticals (where absolutely sterile air is mandatory) or radio broadcasting (where control of sound must be very precise). Even in the skyscraper, it has certain advantages. It makes the wall easier to fabricate, and it eliminates such troublesome elements as movable sash, with its leaks, heat losses, and rattles. For reasons such as these, the sealed wall has been used in many recent skyscrapers.

Nevertheless, the sealed curtain wall—especially in its all-glass version—raised a whole new set of problems for skyscrapers. A simple thing like window washing, difficult enough with movable sash, now becomes a major operation. More fundamental, in these completely sealed and air-conditioned buildings, was the problem of solar heat loads—especially in summer. Even the insulating glasses described above are relatively transparent (62.1 percent) to solar infrared radiation and, contrary to popular belief, shades and blinds inside the glass do nothing to reduce heat; the air conditioning has to absorb the added heat and can only do so indirectly. The simplest solution is to interrupt this radiation *outside* the building skin, as the South Americans do with *soleil brise*. But icing and corrosion have so far prevented the widespread use of this technique, at least in the northern United States. It is for reasons such as these that the designers of many of the most recent curtain walls have felt themselves compelled to return to movable sash and much smaller glass area.

The dominant esthetic problem of the skyscraper has been seen by the architect, from Sullivan's day to the present, as being the "truthful" expression of its peculiar properties. In practice, this has usually meant the expression of its skeleton, and this, of course, has depended upon where the skin or curtain wall was placed. From a purely logical point of view, one would expect it to be always stretched outside the skeleton, since the latter needs the skin's protection almost as much as the enclosed volumes of the building proper. In practice, however, we find the curtain wall located in any one of three vertical planes: [1] outside the skeleton, [2] between the interstices of the skeleton, [3] completely recessed behind the skeleton. Generally speaking, the decision as to where to place the skin seems to be made on a purely esthetic basis. Since it radically alters the external appearance of the building, it is indeed an important esthetic decision.

In the first case, the curtain wall appears as a continuous skin, stretched tight and unbroken around the entire structural frame. The sense of internal articulation is lost. And the odd thing is that glass or metal curtain walls, because of their reflectivity, seem just as opaque as masonry. Thus, for all the actual lightness and thinness of their walls, these buildings appear as solid cubes of glass or solid rods of hammered metal.

In the second case, the curtain wall fills the interstices of the skeleton but is kept on the same vertical plane as its outside edge. Here the esthetic intention is to delineate the skeleton on the façade. How effective this delineation is depends upon the colors and specular qualities of the surfacing materials used. If there is great contrast in color and texture between skeleton and skin, then the delineation is inescapable. If, on the other hand, the skeleton is surfaced in a material similar to the skin, its visual importance will be diminished—may, under certain circumstances, disappear. In any case, it should be noticed that the expression of the skeleton is, at best, diagrammatic and two-dimensional.

Miës van der Rohe is the leading exponent of the third technique—that of depressing the wall plane behind that of the columns. He has done this in Crown Hall at the Illinois Institute of Technology (where the steel frame, being a single story, requires no fireproofing), and in his earlier Lake Shore apartments (where the steel skeleton, of necessity fireproofed, has an armature of flat steel plates bolted on the outside to make the "steeliness" of the skeleton explicit). Where these exposed columns are structural and mark the true bay, the cellular nature of the enclosed volume is apparent. But, where the bay is divided by a number of equally sized, equally spaced verticals, as in his Seagram Building, this nature again disappears. What results is a richly textured surface with a strong vertical emphasis. The building appears as a solid, fluted column.

The reader is free to decide which of the above techniques he prefers as the most satisfactory expression of the structural character of the building. However, it must be observed that all of them share a common esthetic characteristic: they become, ultimately, mere textural manipulations of the surface, like the weave of a fabric. They exist almost independently of the mass or profile of the building as a whole and give scarcely any clue to its internal organization. Moreover, since the panels on a given building are identical, on a big facade they establish a kind of irresistible rhythm which no architect seems able either to interrupt or to halt. Thus the handsomest (though certainly the coldest and most noncommittal) of our curtain-walled skyscrapers are those, like the Lever and the Seagram, where plan, profile, and volume are tailored to fit the modular pattern of the curtain wall. A permissible analogy might be that men's clothing be tailored, not to fit men but rather to fit the woven or printed pattern of the cloth!

How difficult it is to do otherwise, to bend the curtain wall to meet the plastic requirements of the total design, is amply demonstrated in those buildings where it has been tried. Here the architects have attempted to maintain a monumental quality for the building as a visual whole by framing or containing the powerful pattern of the curtain walls. Esthetically, their efforts can scarcely be called successful. In fact, of all recent efforts along these lines, only Frank Lloyd Wright's small skyscraper in Bartlesville, Oklahoma, can be called successful. Here one feels it is the architect, and not the curtain wall, who is in control.

A close examination of this building reveals some possible explanations for its success. In the first place, it is neither regular nor rectangular in plan. Secondly, in addition to office space, part of each floor is given over to apartments, whose more complex space requirements introduce a contrapuntal movement into the composition. Finally, each of the façades of this air-conditioned prism has a different kind of curtain wall designed to handle the climatic conditions of that particular exposure. Wright, with his usual mastery, has converted each of these functional requirements into esthetic assets.

As we have seen, the development of the curtain wall has been umbilically tied to the growth of the skyscraper. And the skyscraper, of all American building types, has been the most generalized and abstract in plan and function. It has housed only one type of commercial activity—office work— and this activity had relatively simple requirements. Moreover, the skyscraper plan developed within the gridiron street pattern, inheriting its rectangularity. This was given a third dimension by the zoning laws requiring stepped setbacks as the building rose in height. All of this led to a building composed of a basic cubical bay which could be indefinitely extended both vertically and horizontally. Thus, esthetically, the paradox of the skyscraper involves the entire structure and not merely the skin alone. Perhaps the way to get more variety and flexibility is to abandon the rectangular plan and the single-purpose occupancy, as Wright did in Oklahoma.

The third aspect of the Wright tower, a different wall for each exposure, shows the most pregnant area of the entire design. Here, for the moment, we must return to purely functional considerations. The typical skyscraper today is a free-standing monolith whose curtain walls are identical on all its façades. This represents a purely formal response to the facts of climate. The buildings are designed as though for an environmental vacuum or—at best—a stable and unchanging set of environmental conditions. In actuality, of course, few climates in the world (and none in the United States) offer anything approaching this state of affairs. Logically, one would expect different types of curtain walls for different exposures and different climates. But this is seldom the case in America—north and south walls, whether in Texas or Chicago, are identical in design.

This sort of unimaginative and inefficient standardization is possible only because of the relative cheapness of fuel and power, as well as of heating and cooling equipment. But it becomes increasingly hard to defend from the points of view of human comfort and mechanical efficiency. Air-conditioning equipment is expected to meet undeviating physiological criteria (e.g., 72° F. air temperature, 50 percent relative humidity) throughout the enclosed volume of the building. Yet around the periphery of this volume, conditions would vary immensely. Thus, on a cold, bright, windy day in December, the north wall—chilled by the wind and untouched by the sun—would have the climate of Canada. At the same time, the south wall of the same building, protected from the wind and exposed to the sun, would have a climate like that of South Carolina. On a hot July afternoon, the west wall would have the climate of the Arizona desert, while at the same time the east wall would have the climate of

Massachusetts. Thus the thermal extremes within which the air conditioning is operating might be more properly expressed in thousands of miles than in tens of feet! Within this continuously shifting pattern of unequal thermal stresses, the air conditioning is expected to maintain a set of stable and uniform conditions.

From a conventional point of view, the most urgent problem in tall buildings is protection from excessive solar radiation in summer. The simplest (though not necessarily the best) is to make the sunny walls, especially the western ones, opaque to solar energy. This alone could reduce the cooling load by as much as a ton of refrigeration for each one hundred square feet. Since many plans do not permit this, architects are increasingly using the *soleil brise* of Le Corbusier and the Brazilians. However, these heavy fixed sunshades become progressively less feasible as one goes north and the length (if not the intensity) of the summer decreases. Hence we find another type of sun screen appearing—a lightweight metal screen mounted out beyond the curtain wall, dense enough in its depth and perforations to exclude most of the high summer sun yet open enough to be unobjectionable in winter. In several of the new buildings this sort of sun screen has been wrapped around the entire building, becoming in effect another specialized membrane of the wall itself. These are not only sensible correctives to the too-transparent wall but also yield some extremely handsome decorative textures to a building type which is all too cold and dull in appearance.

Ultimately, of course, we must demand much more than this of the curtain wall. A technology which can achieve the thermonuclear bomb and the moon rocket should give us a wall which behaves like the epidermis of the animal body—i.e., which responds actively and automatically to changes in its external environment. It is not too difficult to imagine such a wall. In the first place, it should have a capillary heating and cooling system built into it, much like the skin of a warm-blooded mammal. The function of these capillaries would not be actually to heat and cool the interior volumes of the building so much as to provide a thermal symmetry inside which the air conditioning could more effectively operate. A building with such a capillary system would then find its sunny walls cooled with circulating chilled water, even on the coldest winter day, while the solar heat thus picked up would be used by the system to heat the much colder walls on the shaded side of the building.

We can imagine still more efficient and sophisticated building skins than these. For example, in all but polar and subpolar latitudes, enough solar energy falls upon any freestanding building during the course of the year to power that building—i.e., to heat, cool, and light it. The problem, of course, is to trap and store that energy against the hour of need. So far, most solar heat and storage devices are very inefficient, or limited to regions of intense insolation, or both. Though many of these devices could be vastly improved, a new contender, the solar battery, offers interesting possibilites. Assuming that their efficiency could be even modestly increased, the solar batteries might be imagined as forming the outer membrane of sunny walls; they would then pick up sunlight, convert it directly into electrical energy to power the building, storing any surplus of power in conventional storage batteries. Even this system might prove inadequate, however, for the long sunless periods of cloudy climates or high latitudes. If men ever master nature's process of photosynthesis, we might imagine architectural tissue, built on an analogue of the vegetable, which manufactures starch and then stores this energy in the stable form of alcohol for fuel. A range of such possibilities lies theoretically open; by exploiting them intelligently, men might design buildings which would approach the animate world in their operational efficiency.

Of course, some technological breakthrough of a quite higher order may override such developments. For example, if the thermonuclear reaction is finally domesticated, it will supply the energy for a whole new order of environmental control. We can then think of air conditioning entire cities; with such energies at our disposal we could change the climate of whole regions.

Obviously, such developments will radically alter the appearance as well as the structure and performance of our buildings. It need not be for the worse, though the area in architectural design in which personal taste can freely operate will undoubtedly be circumscribed—circumscribed not merely by structural necessities (that has always been the case) but by our vastly increased knowledge of man's physiological and psychological requirements, as well as by the new technological processes he employs to meet them. This in truth will demand a new order of esthetic competence.

III. Episodes in ENGINEERING History

III. Episodes in Engineering History

Although the engineer is today a highly trained and technically competent individual, it must be remembered that he is primarily a practical man, concerned with getting specific jobs done. Sometimes his assignments carry him to the most savage and desolate parts of the earth. Sometimes they call for bravery or skill in diplomacy rivaling those of the soldier or the statesman. The following selection has been chosen because it illustrates in unusual measure some of the most dramatic problems that engineers have been called on to solve. Grenville Dodge, the builder of the Union Pacific Railroad, has been called one of the greatest civil engineers of all time. A general in the Union Army, he made himself invaluable to Halleck, Grant, and Sherman. His surveys, totaling upwards of sixty thousand miles, were indispensable to railroad pathfinding and building throughout the West. His work on the Union Pacific faced almost every conceivable obstacle—political manipulation, financial peculation and speculation, sharply divided corporate management, unruly construction crews, gamblers and cutthroats, the hostility of Indians, and a terrain even more inhospitable than its inhabitants.

BUILDING THE UNION PACIFIC RAILROAD

J. R. PERKINS

SEVEN MONTHS after Dodge discovered the Lone Tree Pass over the Wyoming Black Hills, or on April 24, 1866, he met [Thomas C.] Durant at St. Joseph, Missouri, and held a final conference on the subject of becoming chief engineer of the Union Pacific. The meeting was hardly peaceful. Dodge's old boss, Peter Dey, the first chief engineer of the road, had severed all connection with its construction, being unable to get along with Durant. Dodge was perfectly conversant with all the issues between the two, and he knew Peter Dey to be as high-minded as Thomas Durant was crafty and bellicose.

But there was something else—something far more serious than a squabble over routes out of Omaha: Durant was under fire, both in his own company and in Congress, charged with having built up his personal fortune, through construction contracts, at the expense of the Union Pacific and the Federal government, joint agencies in the building of the road.

"I will become chief engineer only on condition that I be given absolute control in the field," Durant heard Dodge say in his deliberate manner. "I've been in the army long enough to know the disastrous effects of divided commands. You are about to build a railroad through a country that has neither law nor order, and whoever heads the work as chief engineer must be backed up. There must be no divided interests; no independent heads out West, and no railroad masters in New York."

Whatever Thomas Durant may have thought and felt, he was far too shrewd to oppose the ideas of the man that both Grant and Sherman wanted to become chief engineer of the Union Pacific. Besides, Durant had the good sense to know that Dodge's initiative, experience as a soldier and training as an engineer made him the one man for the position. So the bargain was struck and Dodge returned to his command at Fort Leavenworth and wired Grant for a leave of absence from the army, for Durant told him that it was imperative that he come to Omaha at once and assume charge of engineering affairs.

It was on May 6th that Dodge entered the chief engineer's office in the second story of a little brick building occupied by the United States Bank at Omaha. Disorganization was apparent, for there was no regular head to the company west of the Missouri River, and the engineering, the construction and the operating departments were all reporting separately to New York. And in New York City was a little group of railroad promoters who knew nothing of building across the plains and who, as a consequence, were quarreling among themselves.

The Union Pacific Railroad Company, at this hour, may be likened to that individual in the popular song who said, "I don't know where I'm going but I'm on my way." In other words, the final surveys for the road were never far ahead of its actual construction, and when Dodge assumed charge no one knew whether it would be built out the North Fork Platte toward Fort Laramie; out the South Fork Platte to Denver; or due west from where the Platte divides, out Lodge Pole Creek. And as to building across the mountains after traversing the plains, there was neither agreement nor understanding. Isolated engineering parties were roaming both the plains and the mountains, and had been for upward of twelve months; but, decimated by the Indians and discouraged by receiving no pay, some of them had disbanded and others sat down to await developments.

One may keep well within the bounds of truth and say that General Dodge knew more of the possibilities of the country from the Missouri River to Salt Lake, from a railroad standpoint, than any other American engineer. He was the first engineer to be employed by a railroad company to make surveys out the Platte River Valley and on to Salt Lake. He had been sent by Henry Farnam of the Rock Island ten years before the Union Pacific drove its first spike at Omaha, and his surveys were something more than horseback reconnaissances, for he had used his instruments and the Indians had named him "Long Eye."

But nothing is more difficult in the history of the first transcontinental railroad than to determine the value of surveys and the place of surveyors. First, there were the buffalo trails, and no one knows how old they were before the Indians rode them; there were the emigrant routes superimposed on the Indian trails; there were the routes of the overland mails superimposed on both; and then came the builders of railroads. As Dodge said:

> There was never any very great question, from an engineering point of view, where the line, crossing Iowa and going west from the Missouri river, should be placed. The Lord had so constructed the country that any engineer who failed to take advantage of the great open road out the Platte valley, and then on to Salt Lake, would not have been fit to belong to the profession.

In thirty days Dodge completed his organization, and on a military basis. Isolated engineering parties, scattered from Fort Kearny to Salt Lake, were brought into coordination, provided with heavily armed escorts and ordered to swing into action again; construction parties, long idle, were stimulated to activity; materials began to move, and the thud of the sledges on the spikes told of the new beginning.

Three months after Dodge became chief engineer, Jack Casement assembled a thousand men and one hundred teams out on the prairies of Nebraska, forty miles from Omaha, and told them what he expected. It was a mixed crowd of ex-Confederate and Federal soldiers, mule-skinners, Mexicans, New York Irish, bushwhackers, and ex-convicts from the older prisons of the East. Somewhere in California was another group pushing the Central Pacific eastward.

"Boys," Dodge said, "I want you to do just what Jack Casement tells you to do. We've got to beat that Central Pacific crowd."

With a wild raw yell Casement's men swung into action and the track-laying of the Union Pacific increased to three miles a day within a month. The East heard of it and out came bankers, statesmen, magazine writers, and special correspondents. The Union Pacific Railroad, a dream, a theory in the opinion of many, was lifting to reality. And this is what they saw:

Long lines of grading teams sinking scrapers into the soil of Nebraska where plows had never gone before; great wagon trains of ties rolling in from the Far West; a hundred bronzed men dropping the timbers in their place; a hundred others pulling iron rails from flat-cars and dumping them along the embankment; a dozen brawny Irishmen tugging the rails to their position; the falling of the sledges; the rhythmic bending of the bolters; the laying of four rails to the minute; and the steady creeping, like a great brown worm, of the track to the west.

But this first group of easterners saw something else: they saw the beginning of the "moving town," of the "hell on wheels," for at each base, in increasing ratio, there assembled the strumpets, the gamblers, the liquor dispensers of a dozen states; and Paddies, troopers, Indians, and frontiersmen drank, sang, danced, gambled, and fought.

The first moving town sprang up at Fort Kearny on the Platte River, the second at North Platte, and the third at Julesburg, which turned out to be the worst of the three. The rougher element figured that Julesburg was far enough west to be beyond the pale of law and order. A group of gamblers took possession of the town as soon as Casement and his crew began to work west of it, jumped the land that Dodge had set aside for shops and defied all creation. Dodge ordered Casement to return to Julesburg and restore order. Three weeks later the chief engineer had occasion to visit the town.

"Are the gamblers quiet and behaving?" he inquired of Jack Casement.

"You bet they are, General," Casement replied. "They're out there in the graveyard."

He had descended on the town with a hundred seasoned soldiers and wiped out the ringleaders.

It was in August, 1866, that the Indians first began their attack on the builders of the Union Pacific. The road had reached Plum Creek, two hundred miles west of Omaha, when a powerful band of Indians swept down on one of the freight crews, captured it, and held the train. The situation was acute from more angles than one. Dodge had no hesitancy in cleaning out troublesome whites, but the Indians, being wards of the government and the subject of much emotional oratory, presented another problem.

He was ten miles west at the end of the line when word came to him of the trouble. He ordered his private car—an arsenal on wheels—hooked on to the handiest engine and with twenty men raced back to the scene of the capture. Halting his train he deployed his men on either side of the track and opened fire on the Indians, who had set fire to the freight. They mounted their ponies and ran without making a show of resistance. It wasn't much of a fight, but it marked the beginning of twenty months of bitter warfare against the building of the first transcontinental railroad.

But more than Indian troubles now confronted the chief engineer: there were bitter disputes between the government directors and the Union Pacific officials over the route of the road through the mountains. Some favored one pass and some another, but the chief engineer had plans of his own. He headed straight for the lone tree that marked the defile he had discovered the year before when beset by Indians.

A thorough examination of both the eastern and the western slopes of the Wyoming Black Hills convinced Dodge that the road should be built through the pass he had discovered and he instructed his engineers to make the exact location. His decision marked the beginning of his disputes with Durant over the location of the line from the eastern slopes of the Black Hills to Green River, far to the west—disputes that all but halted the building of the road.

On October 8th, Dodge and escort plunged into Boulder Canyon to hold a conference with one of his engineering parties when a heavy snowstorm swept down on them and caught them in a critical situation. The teams refused to face the snow that was fast turning to a stinging sleet. Dodge ordered the packs to be taken off the mules to let the animals shift for themselves, and after a hard struggle he succeeded in leading his men to the shelter of an old stamp mill near a mine. For three days they were snowed in and provision ran low, and Dodge saw enough of Boulder Pass to convince him that the main line of the Union Pacific could never be built any closer to Denver than a point a hundred miles to the north. While cooped up in the mountain storm he was elected to Congress, having been drafted by the soldier element of Iowa early in the spring to oppose John Kasson for the Republican nomination. "But I'd even forgot," Dodge said, "that it was election day."

By the middle of October Dodge was back at headquarters in Omaha. A letter from Thomas C. Durant conveyed the information that the vice-president of the Union Pacific was about to lead a large party of eastern people to the prairies of Nebraska to see the railroad. Durant was short of money and he planned the trip to interest certain well-known capitalists from New York, Boston, and Philadelphia.

Dodge got in touch with the quartermaster general at Omaha, secured a great number of tents and set up a first camp out on the Loup Fork River. He sent for Major North, who had commanded the Pawnees in the Indian campaign of 1865, and told him to select fifty of his best red men, put them in their war paint and camp them on the opposite side of the river under cover of the darkness and await orders.

A week later, Durant arrived in Omaha with "one hundred fifty prominent citizens and capitalists and ladies to see the railroad and the country."

When Durant and party arrived in Omaha their equipment resembled an old-fashioned traveling show. There were a band, a caterer, six cooks, a photographer, three "tonsorial artists," a "sleight-of-hand performer," and a printing press. They had provisions enough to feed a regiment and of a kind that a regiment never ate.

From Omaha the excursionists went out over the new Union Pacific Railroad to the camp Dodge had prepared on the Loup Fork River, and, after a big supper, a bonfire dance and a musical program, they retired to their tents. At three o'clock the next morning Dodge took an engine and crossed the river to where Major North was camped with his Pawnee scouts, loaded them on the coal-tender, the pilot, and wherever an Indian could cling, and at dawn backed across the river to the camp of sleeping easterners.

The Pawnees, following instructions to the letter, stole into the camp fully dressed in their war trappings and began to whoop at the top of their voices. The surprise was so complete that for the next minute great excitement prevailed and a couple of the ladies found it convenient to faint. But the whole affair ended in a friendly dance around the fire followed by breakfast, and the Indians left the camp ladened with gifts.

Dodge conducted the excursionists on to the end of the road, and there were hunting parties in which buffalo and antelope were rounded up in droves so great that the amateur hunters could not miss them, and many Union Pacific bonds were negotiated over the fires that roasted the meat, just as the shrewd vice-president of the road had planned.

Dodge went east after the excursion was over, presented his plans for building the Union Pacific through the mountains, secured the consent of the directors to construct over the Lone Tree divide, and then hurried back to the West. More difficulties than he had ever dreamed of were at hand, for Durant was beginning his struggle with the Ames brothers for the control of the road; moreover, the Indians were fully organized to give their best blows.

Dodge wrote Sherman a long letter, told of his plans to reach Fort Sanders—288 miles west—in another twelve months, and frankly declared that he must have more troops and more authority.

Grant and Sherman moved at once and General Augur was given command of the Department of the Platte and told to cooperate with Dodge in the building of the Union Pacific Railroad. Augur came and asked Dodge what he wanted and Dodge replied:

"I want strong military escorts with each party of engineers and I want detachments strung all along the line from Alkali to the Laramie River."

During the winter and the spring Dodge fought snows and floods. In March tracks were blocked, for no one had anticipated snow blockades. In April great floods swept through western Nebraska, tore out miles of track, bridges, and telegraph poles. The damage at the Loup Fork bridge alone exceeded fifty-thousand dollars. And just at this inopportune time the company sent its representatives west to see what Dodge was doing. No visit could have been more inauspicious. Moreover, Durant and Dodge were quarreling. Dodge was frank with Oliver Ames, the president of the road, who came out.

"Durant is in the way," he told Ames.

Oliver Ames sized up the situation and sat down and wrote:

"It shall be the duty of the chief engineer of the Union Pacific Railroad to take charge of all matter pertaining to the construction of the road."

The floods of nature and the disputes of human nature had been unable to halt the chief engineer of the Union Pacific, so the Indians took a hand. From the deep ravine of the Wyoming Black Hills they swept down to Lodge Pole Creek, pulled up the stakes that marked the line of the road, stole the teams and drove the workmen back upon the base; they struck another party on the Laramie Plains, cleaned it out and burned everything in sight; and they wound up by tackling one of the best protected engineering groups on the road; killed a soldier and a tie-hauler, and playfully burned the stage stations along a fifty-mile front.

Dodge, greatly troubled if not discouraged for the first time, wrote Sherman on May 20th: "I am beginning to have serious doubts of General Augur's ability to make a campaign into the Powder River country and at the same time give ample protection to the railroad, the mail routes and the telegraph."

The last week in May three government commissioners, White, Simpson, and Frank P. Blair came west and went to the end of the track to examine the road. They had just completed their task and were standing talking with Dodge when more than one hundred Indians suddenly swept down a ravine and made a fierce assault upon the workmen at their lunch. A company of soldiers was less than a mile away, but they were unable to render any assistance, so swift was the rush of the red men. The whine of the bullets uncomfortably close to the government commissioners, the desperation of the attack, and the indifference of the Indians to their own fate convinced Blair and the others that the red men were now fighting for their country and not merely to steal a few head of cattle.

Dodge left the commissioners standing on the hill, jerked his revolver, and hurried down to the tracks, yelling at the chief graders to grab their guns and go after the Indians, but the workmen were demoralized and sought the shelter of the freight cars and would not budge. He stormed about, upbraided the graders for allowing a band of savages to rush them without firing a shot in return, and then went back to the commissioners and tartly said:

"We've got to clean the damn Indians out or give up building the Union Pacific Railroad. The government may take its choice."

He was despondent, ill, and had begun to feel that he stood alone in his efforts to complete the Union Pacific in the time allowed under the Act of 1862. But Blair told him that he had seen enough to be convinced that the road could not be pushed through the mountains without heavy re-enforcements from the army and he promised to go back east and stir up things. He did so and three additional companies of cavalry were stationed along the line for scouting purposes, and to keep the Sioux and the Cheyennes pushed back into their own country.

The parties pushed out Crow Creek to the new railroad base, and ten thousand people, alert to every advance the road made, poured into Cheyenne in advance of the track-laying. It was like a gold rush; they milled through the new streets, quarreled and fought over town lots, put up gambling houses and saloons as well as shacks to live in, and clamored for the railroad company to establish its shops there.

"The company will do nothing of the kind unless there is more law and order here than we had at Julesburg," Dodge made known.

White adventurers were about to disorganize his construction crews and he threatened them with the military in the new post hard by. A remnant of the old Julesburg crowd that Jack Casement had cleaned out decided to be good—as good as they had to be—and a city government was organized.

But the Cheyenne Indians up Crow Creek, ignoring the compliment Dodge attempted to pay when he named the new town after their tribe, hovered on the outskirts and waited for an opportunity to strike. Cheyenne stood at the gateway to their mountain fastness, and they had no intention of standing idly by and seeing a railroad—which meant white supremacy—tap that very territory. So they watched Cheyenne's first Fourth of July celebration from afar, heard the reverberation of "anvil-shooting," heard, though faintly, the playing of the band, and saw the shooting of rockets when night came.

On the morning of July 5th, when not a few of the celebrants were sleeping off the effects of whisky, the Indians slipped down the hills, rushed a party of graders, killed a few, stole the horses, circled the town and, yelling like mad, vanished up Cheyenne Pass.

Two weeks later Dodge and his party started on the long march to Salt Lake. The escort toiled to the summit of what is now known as Sherman Pass, 8200 feet above sea level, and descended to Dale Creek, the most serious obstacle to the building of the Union Pacific, for this stream required a bridge 125 feet high and 1400 feet long.

Fort Sanders was reached in a few days, and there Dodge met General Gibbons, one of Grant's most trusted officers of the Civil War. Nothing could have been more fortunate in the building of the Union Pacific from this fort west to Green River than the presence of General Gibbons, who entered sympathetically into all of Dodge's problems. He was in conference with Gibbons when a rider, with foaming horse, dashed up to the post with the news that the Percy T. Brown engineering party, engaged in the difficult work of making locations across the Great Divide, had been severely beaten by the Sioux, and Brown had been killed.

Dodge bowed his head and groaned, for not only was

Percy Brown a capable engineer but he was devoted to his chief and trusted above anyone who had ever operated between Rattlesnake Pass and Green River. Then Dodge heard the story. Brown and thirteen men, beset by a powerful band of three hundred Sioux Indians, fortified themselves on an elevation in the Great Basin and fought from noon until night, when their foes, for reasons unknown, withdrew. Brown, badly wounded, begged his men to leave him, but they refused, made a litter of their carbines, and carried him twelve miles to a stage station, where he died within an hour.

Dodge pushed on west to strengthen his badly disorganized engineering parties all along the line. At Rattlesnake Pass he discovered coal and later sank a mine, which was the first that ever supplied the Union Pacific along its own route. On reaching the North Platte he found it to be swollen from heavy mountain snows. He ordered it to be forded, but two of the young officers who attempted it were swept back to their starting point, and the remainder of his escort refused to budge.

Dodge jumped on his own horse, Rocky Mountain, and plunged into the stream, calling on his entire command to follow. It proved to be a desperate undertaking and three of his men had narrow escapes from drowning.

"If you are going to help me build the Union Pacific through this country you've got to learn to swim horses across more rivers than this one," the chief engineer said.

Near this crossing of the North Platte he and General Gibbons established Fort Steele, for just to the south was the Medicine Bow range, rendezvous of the Crows, the Sioux, and the Cheyennes, and no railroad could be built until they were held back.

They came upon the Percy Brown engineering party, strongly reenforced it, started it to work and then pushed toward the Great Divide. At their feet was a vast basin. Dodge lifted powerful glasses and looked in every direction.

"I see some teams down there. Must be white men—perhaps returning emigrants."

He plunged down into the basin, followed by his escort,

and an hour later came upon one of his own engineering parties—the one headed by Charles Bates that had been ordered to survey from Green River back east to meet Brown. Bates and his men were in bad condition, without water, and with swollen tongues. But they were at work, running a true line, for they were the kind of men who made the Union Pacific possible. Colonel Steptoe had attempted to cross this portion of the basin on his first trip to Oregon, but he had had to turn back. Later, Dodge came upon old broken wagons, anvils, and other tools that Steptoe had abandoned.

It was late in August when the party reached Fort Bridger, but the famous guide was on a scouting trip. He and Dodge were close friends and when Bridger died in 1881 Dodge launched a movement that resulted in a monument being erected to him in a Kansas City cemetery.

From Fort Bridger they pushed on rapidly toward Salt Lake. Dodge's party reached Salt Lake the afternoon of the thirtieth and camped south of town. They had scarcely settled when out came Brigham Young and his favorite wife, Amelia Folsom, who was from Dodge's home town, Council Bluffs. The head of the Mormon church was quite agreeable. He was watching all surveys, even participating in a few, for he wanted to be sure that the first transcontinental railroad touched Salt Lake.

On September 4th, Dodge led his command out of Salt Lake for the 700-mile trip back to Fort Sanders. He had it in mind to examine the approaches to Snake River with the view of running a line from Promontory Point in Utah to Puget Sound and thus affording the Union Pacific a tidewater outlet, for it had slowly dawned on him and on every one else connected with the road that the Central Pacific would block the way to San Francisco. Moreover, he learned to his own satisfaction that the valley of the Snake River would afford the Union Pacific its most feasible route to Idaho, Oregon, and Washington.

On arriving at Fort Sanders, Dodge made preparations to go to Washington and take up his Congressional duties.

He entered Congress almost at the beginning of the legislation that was hostile to the Union Pacific, and the railroad company, whether he did or not, always counted it most fortunate that he was a member of the Fortieth Congress. But not for a day did he relinquish his hold on the railroad situation in the West. From his office in the Department of the Interior, especially placed at his disposal for his double task, he sent out his orders to the division engineers who, in turn, kept him posted on all developments. And there were many developments!

Near the close of the session of the Fortieth Congress word came from the West that Durant, in violation of every agreement made with Dodge when he became chief engineer, was changing locations, altering the line over the Wyoming Black Hills, and circulating rumors that the Union Pacific Company would make Laramie City, and not Cheyenne, the mountain division of the road. Moreover, the construction crews between Cheyenne and the new terminus town of Laramie Ctiy were threatened with extinction by a powerful gang of gamblers and whisky venders operating in a half-dozen camps.

When Dodge received this news he drew down the top of his desk, packed his valise and caught the first train out of Washington for the West. On reaching Cheyenne, he found a muddled state of affairs. Scores had left the town, trekking across the mountains to the new base at Laramie City which, according to Durant, would soon have the shops of the Union Pacific Railroad. The citizens of Cheyenne—those who had invested heavily in town lots—called on Dodge and demanded an explanation.

"You can't prevent an exodus from an old base town to a new one. You fellows ought to know that," Dodge said. "But on the score of the removal of the shops, well, Durant lied. The shops will remain in Cheyenne; the branch line will be built from Denver, and this will be the division."

Dodge ordered his private car hitched to an engine, wired for the track to be cleared and raced west across the mountains toward Laramie City, then the end of the line. It squatted on the western slope of the Wyoming Black Hills and felt its importance, for it was, in May, 1868, the end of the freight and passenger division and the beginning of all the construction work west to Green River. The same crowds that had flocked from Fort Kearny to Julesburg and from Julesburg to Cheyenne had poured into Laramie City, and they had grown wilder as they progressed.

The "Big Tent" was up and doing a thriving business the evening Dodge arrived. It was the town's social and civic center and it was just a little bigger and a little tougher than it had been at the other points. From a platform a German band played noisily; and while the mule-whackers, miners, and railroad workers danced with the strumpets, scores of others crowded the gambling tables, played monte, faro and rondo coolo; and against the long bar, with its background of cut-glass goblets, ice pitchers and high mirrors, leaned those who drank hard whisky and sang the sentimental songs of their childhood back in the older states.

Dodge's visit to Laramie City was a dash of cold water.

"The shops will remain at Cheyenne," he said. "And if the gamblers and saloon-keepers here don't let the railroad employees alone, I'll have General Gibbons send down a company of soldiers and we'll proclaim martial law. Take your choice."

Then he hunted up Thomas C. Durant, and the meeting was far from pleasant. Durant was making a final bid for power and authority, for he was at war with the Ames brothers who were pushing him steadily into the background. But Durant held as much stock as anyone, and other directors of the company feared him. He had used Dodge in pushing the Union Pacific across the mountains, but he now believed that he could get along without him; indeed, he felt that Dodge was in his way, and he planned to elevate Colonel Seymour, consulting engineer, to the position of chief engineer.

"Durant," Dodge said in his deliberate way, "you are now going to learn that the men working for the Union Pacific will take orders from me and not from you. If you interfere there will be trouble—trouble from the government, from the army, and from the men themselves."

Dodge turned abruptly and left Durant standing in the

dusty Main Street of Laramie City, and the rails of the Union Pacific began to be laid faster than ever before. Thousands of emigrants rolled along the trails in covered wagons; the engineers, obeying Dodge to the letter, linked up their lines from rivers to mountains; the road crossed the North Platte and pushed into the desert; the town of Benton—the last wild terminus of the road—was born; the grasshoppers came in great armies and ruined the crops of the settlers; flour jumped to eight dollars a hundred pounds; the Cheyennes struck savagely at a dozen points, killing and scalping graders and even the crews of freight trains; but the chief engineer of the Union Pacific, dominant and in his full powers, raced from one end of the track to the other and drove unceasingly.

With everything moving along the line of the Union Pacific, Dodge took a strong escort, left Benton the first week in July and pushed west to bolster the morale of the advanced engineering parties. The journey is the record of a man whose whole powers were fully dedicated to the task of pushing to completion the first great transcontinental railroad.

On July 23rd, while camped near Salt Lake, Dodge received a telegram from Sidney Dillon, a director of the Union Pacific, requesting him to return to Fort Sanders with all possible speed to meet Grant and Sherman. The dispatch also carried the information that Durant had, in Dodge's absence, secured larger powers from the company and would stand on that authority in the conference to be held with government commissioners and military heads en route to Fort Sanders from the Sherman-Harney Indian treaty at Fort Laramie.

Dodge staged it back to Benton as fast as he could and then caught a train for Fort Sanders, where he met his old commanders. With them were Generals Sheridan, Harney, Kautz, Potter, Dent, Hunt, and Slemmer. Grant had made up his mind to take a hand in the affairs of the Union Pacific, for the troubles of the company had aroused the authorities at Washington to the seriousness of the situation.

It was on Sunday, July 26, 1868, that this notable group, augmented by Sidney Dillon, Thomas Durant, Colonel Seymour, and Jesse L. Williams, a government commissioner, met the chief engineer of the Union Pacific in a conference that had marked bearing on the building of the final 600 miles of the road.

Durant, believing that his new powers with the Union Pacific Company gave him undisputed authority, took the floor and boldly charged the chief engineer with having selected impossible routes, wasted money in useless experiments, ignored the sound judgment of his associates and failed to locate the line as far as Salt Lake.

"What about it, Dodge?" General Grant inquired, leaning back in a cane-bottom chair and smoking vigorously.

"Just this," Dodge began deliberately, "if Durant, or anybody connected with the Union Pacific, or anybody connected with the government changes my lines I'll quit the road."

There was a tense pause; Grant shifted his cigar, Sherman's seamy face was immobile, but the others were ill at ease. Durant's delicate fingers pulled at his Van Dyke beard; he glanced at Colonel Seymour, his henchman, but he said nothing. Grant finally broke the silence.

"The government expects this railroad to be finished," he said slowly. "The government expects the railroad company to meet its obligations. And the government expects General Dodge to remain with the road as its chief engineer until it is completed."

It was a dramatic moment; it was even a critical moment in the building of the first great transcontinental road. Durant looked at the man who would soon become president and doubtless did some quick thinking. Anyhow, whatever he thought, he turned to Dodge and said:

"I withdraw my objections. We all want Dodge to stay with the road."

An hour later the entire group posed for a local photographer, little dreaming how significant the picture would be for another generation.

"Engineering is the combination of art and science by which material and power are made useful to mankind," says the Engineers' Council for Professional Development in Engineering as a Career. An outstanding example of this marriage between art and science is the Brooklyn Bridge, one of the world's great engineering landmarks.

The building of the Brooklyn Bridge is intimately connected with the story of a father and son, one of whom sacrificed his life and the other his health in its construction. John A. Roebling, the inventor of wire rope, made the original design for the bridge and in spite of strong opposition succeeded in having it adopted. In the early stages, his foot was crushed by a boat coming in to dock. Tetanus developed and he died soon afterward. His son Washington, who had been trained by his father, took up the work with a zeal and dedication which were almost religious. He too fell a victim. While working with an underwater crew, he developed caisson disease, now known as the bends. He became paralyzed but continued to direct construction from a room near the bridge, observing details with field glasses and relaying his orders by messenger. Every step toward completion was a tribute to his courage and engineering skill. The official dedication on May 24, 1883 was recognized as an event of world importance.

The whole story is told in The Building of the Bridge by David Steinman, from which the following episode is here reprinted. Steinman is not only the author of a number of engineering books, both popular and technical, but is also a worthy successor to the Roeblings. A Professor of Civil Engineering, he was responsible for many innovations in bridge building. Among others, he designed the Triboro and Henry Hudson Bridges in New York and the Thousand Islands Bridge across the St. Lawrence River.

THE CAISSONS OF THE BROOKLYN BRIDGE

D. B. STEINMAN

WITH ITS INSPIRATION GONE, the Brooklyn Bridge seemed impossible to build, for John Roebling had been more than the designer. His had been the vision, the conception, the knowledge of the whole and of every part, the anticipation of every problem of execution. A new guiding spirit was needed to carry the great work forward.

By destiny or prevision the master builder had prepared a deputy who was indeed a part of himself—his own son, Colonel Washington A. Roebling. This young but able engineer had been trained and prepared for this task. He had been given the best engineering education of the time; he had received his practical training under the greatest bridge-builder of his day—his own father; and his own judgment, initiative, and courage had been tried and tested on the battlefield.

When the elder Roebling began planning the Brooklyn Bridge, the son took an immediate interest in the great adventure. He helped his father in the preliminary studies; and then he journeyed to Europe, visiting all the important engineering and metallurgical works and studying caisson work and the fabrication of the new structural material, steel. With the mass of data and information he had collected, he returned to America to serve as his father's aide for twelve months before the latter's death. He was thoroughly familiar with the plans for the bridge, having helped to create them.

To this son the great engineer on his deathbed turned over the completion of the task. The following month—August, 1869—at a meeting of the directors Colonel Washington A. Roebling was officially appointed to succeed his father as Chief Engineer of the work.

In the approximately thirteen years from the actual beginning of the work, when laborers started to clear away the site for the Brooklyn tower on January 2, 1870, to the opening of the structure to the public on May 24, 1883, deaths and casualties were many. Into the construction of the monumental span went blood, sweat, and anguish in equal measure with cement, stone, steel, and treasure.

The first problem was the construction of the foundations for the two massive piers in the river. In order to carry these granite masses to their towering height, a firm foundation was indispensable. For this work pneumatic caissons were adopted. This was then a new method of building deep foundations under water, just being introduced into this country and involving many unknown elements of difficulty and hazard.

A pneumatic caisson for sinking subaqueous foundations is merely a diving bell on a vast scale. It is essentially a huge airtight box, having a roof and sides but no bottom, in which men can work under the protection of the compressed air which keeps the water from entering and filling the chamber. The "sandhogs," as the men working within the confined space are called, handle the excavating of the material below them so as to permit a regulated sinking of the caisson under the superimposed weight of the masonry which is simultaneously being built up in successive courses upon its roof.

At the Brooklyn tower site a trial boring, made by the elder Roebling in 1867, showed solid rock at a depth of over ninety feet below high water. To carry the foundation down to that depth would have added enormously to the cost. Fortunately this was not necessary. At a depth of about forty feet the material was so hard and compact as to be entirely satisfactory as a foundation.

It was necessary, however, to provide a uniform distribution of the load over the entire area, whatever the depth might be, so as to insure an even settling from the inevitable, though slight, compression of the underlying material. The magnitude of this aspect of the problem became evident when it was considered that the base had to support a weight of 80,000 tons, in the form of a masonry tower extending to a height of more than three hundred feet above the foundation.

Special consideration had to be given to the design of the roof of the caisson for adequate strength and rigidity, since it would have to carry the superincumbent mass of masonry during the sinking and would furthermore have to serve as a permanent foundation platform to carry the full load of the tower after completion.

Colonel Roebling decided that a solid timber platform of sufficient thickness to act as a beam for carrying and distributing the superimposed load would meet the requirements of the case. He knew that timber permanently immersed in water is imperishable, and if sunk beneath the river bed would be safe from the ravages of sea worms. This massive timber platform would constitute the roof of the caisson and would later remain as a permanent part of the foundation under the tower. The buoyancy of the timber would moreover facilitate the lowering of the caisson without the use of powerful and expensive mechanical rigging.

Work was begun on the caisson for the Brooklyn tower. The roof was made a solid mass of timber, initially five feet thick, to be subsequently increased to fifteen feet to support the weight above. The walls, also sturdily fashioned of timber, were made nine feet thick at the top and tapered to form a cutting edge at the bottom. This cutting edge, or shoe, to penetrate the sand and clay, was shod with a heavy rounded iron casting and was armored with boiler plate.

To make the caisson airtight, the seams were all thoroughly calked with oakum inside and out. In addition, to protect the timber from the teredo and other marine borers, an unbroken sheet of tin extended over the whole caisson between the fourth and fifth courses of the roof and down the four sides to the shoe.

The caisson was constructed close to the Brooklyn shore at Greenpoint, on seven launching ways, sloping down into the water. It was built like a huge inverted flatboat, and with the broad side toward the water. The two end walls and the five intermediate partitions framed inside the caisson provided direct bearing on the seven supporting ways. Full complements of wheelbarrows, crabs, winches and other tools were placed within the caisson for future use. The total launching weight was three thousand tons.

On March 19, 1870, the launching took place. It was considered a noteworthy feat of engineering and a success in every respect. As soon as the last block was split out and the control cams were released, the caisson began to move. The impetus it acquired in the first part of its course proved sufficient to overcome the immense resistance offered when it struck the water. Exactly as planned the air caught inside of the air chamber assisted materially in buoying up the mass. The deck was not submerged, nor was a large wave formed in front.

An air pump and boiler which had been set up on deck before the launching were promptly put in operation, and in a few hours the water was all displaced from the air chamber, the surplus air blowing out at one corner. This provided proof that a satisfactory state of airtightness had been secured. Afterward, when the air was all allowed to escape, the top of the caisson settled down to within seventeen inches of the surface of the water, which agreed with previous calculations of its buoyancy.

In the meantime the permanent site was being made ready to receive the caisson. This was the old ferry slip where John Roebling had received his fatal injury. The preparation of the site consisted in establishing a rectangular basin, open on the side toward the river, the remaining three sides enclosed by a wall of timber sheet piling, and the bottom leveled by dredging to a uniform depth of eighteen feet below high water, a depth sufficient to float the caisson at all stages of the tide.

The dismantling of the old Fulton Ferry slip—pulling out piles, tearing out fender sheathing, removing heavy cribwork filled with stone, and dredging off the loose material on top—required about one month's labor.

All the old timber and piles taken out were found to be infested with thousands of sea worms; their ravages, however, were confined to the space between low water and the mud line. A pile which was sixteen inches in diameter below the mud line, and perfectly sound and free from worms, would be eaten away to a thin stem of three inches just above, all timber being affected alike. This experience showed the necessity of going below the river bed with the timber foundation, and also proved its entire safety in that position.

After about ten thousand yards of old fill and surface mud had been removed, a line of soundings was taken that showed three thousand cubic yards yet to be removed before the level of eighteen feet would be reached. The character of this material was next to solid rock; none of the dredges could make the slightest impression upon it. In fact, all the old harbor charts showed this point as a reef of rock. Under these circumstances recourse was had to powder. Holes were first made in the bottom, by driving down a pointed pile and then withdrawing it. Three such piles were used, shod with iron at point and head. Into each hole a canister of powder was inserted by a diver, and the charge was then exploded by electricity.

After about one week's work the canisters were made heavier, using thick shells of cast iron. These possessed the advantage of dropping to the bottom of the hole by their own weight. After a thorough blasting the dredge could work to advantage for a time. Boulders too large for the dredge to handle were slung with wire rope under water by divers, and either raised to the surface or floated under water beyond the enclosure.

The casual observer above, to whom the surface of the water appeared the same, day after day, would think that nothing was being done. But the divers who slung boulder upon boulder, night after night, had a different story to tell.

The driving of the iron-shod piles afforded a thorough knowledge of the entire ground. On the easterly side a few blows would force the pile through soft clay. In the center, however, there was a broad ridge of hardpan of varying thickness. It was a compact mass of sand, clay, and boulders cemented together so hard that frequently a hundred blows of a 1500-pound hammer were required to drive the pile three feet into the material.

While the dredging progressed, the enclosure proceeded. An outer row of piles was first driven, and anchored back to resist a high bank of fill. Within this line a row of timber sheet piling was driven, space being allowed to tow in the caisson.

On May 1st, the site having been sufficiently leveled, the strange upside-down craft was towed down the river from Greenpoint, a distance of five miles, by six tugboats, under the command of Captain Maginn. During the trip the air chamber had to be kept fully inflated, as for part of the journey there was but a foot of space between the river bottom and the lower edge of the caisson. On the following day the caisson was warped into position and carefully centered with the aid of surveying instruments. By June 20th the ten additional courses of roof timber, to complete the fifteen, were laid.

As the roof of the caisson was built up, additional sections of the large wrought-iron shafts were put in. These included two "water shafts" for taking out excavated material, two "man shafts" for entering and leaving the caisson, and two "supply shafts" for sending down supplies during the sinking and for delivering the material to fill the caisson when it was sunk to its final level. There were also a number of smaller pipes passing through the roof, for supplying gas and water and for blowing out sand by air pressure.

All of the doors to the air locks and shafts fitted closely, and were mounted to swing downward into the space or chamber having the greater air pressure.

On May 10th the air chamber was first entered by the engineers and workmen, and explored. Removing the temporary wooden bulkhead used in the launching, clearing away boulders and all loose material from under the edges, and cutting doorways through the main division frames occupied several weeks. During this period the caisson was rising and falling with every tide, and not until three courses of masonry had been laid was it weighted sufficiently to rest firmly on the bottom and resist the action of the tides, its top still visible above the water. Now the operation of sinking the caisson could proceed. By this time the force of men had been increased to over one hundred, and the work could be carried on continuously. The men worked in three shifts of eight hours each.

The work of excavation was carried on from the air chamber. Any obstructions encountered under the shoes and under the partition frames were removed to permit the sinking to progress. At the same time the masonry was being laid on top, with the aid of boom derricks and hoisting engines.

Crossing the East River by ferry, passengers noticed an area of strange activity near the Brooklyn shore—puffs of steam engines, the clanking of chains, and the groans of huge derricks. Squads of workmen were hustling about their tasks on a mounting pier of masonry. Immense blocks of stone were being landed from a schooner, hoisted, and set in position. But the most fascinating part of the work was invisible, the execution of a great engineering feat deep below the bed of the river.

Within the caisson the laborers, with pick, shovel, and barrow, kept feeding the dredges with the mud and stones they dislodged. When the earth had been sufficiently removed, the workmen drove out alternate wedge blockings from under the partitions on which the mass rested, leaving its weight on the remaining blocks; these slowly settled under the load, and in the most uniform manner. The earth was leveled, and the wedges previously removed were again tightly driven in. Then the other wedges were driven out, and in the whole operation the caisson had gone down a fraction of an inch. The weight and strength of the caisson and the load it carried were so enormous that any difference in the density of the bottom material or in the driving of the wedge blockings was irresistibly equalized, and the mass sank, inch by inch as the process was repeated, plumb down. Of course care was perpetually taken that no boulder or other obstruction should be left under the descending shoe of the caisson's edges, and no pains were spared to remove any cause that might tend to deflect the perpendicularity of the descending mass.

But for all that, the work was anything but easy. Often the material to be excavated was so hard and tough that the picks and shovels could not loosen it, and steel bars specially prepared had to be driven in and small portions at a time picked off. Boulders kept turning up in the most unfortunate places, in all sizes up to ten cubic yards, most frequently of traprock and sometimes of quartz or gneiss. They represented a complete series of the rocks found for a hundred miles to the north and northeast of Brooklyn. These boulders had been transported and deposited there by the ancient glacier. Men in the working chamber constantly probed under the water for these hidden obstructions, in order to prevent the caisson from settling on them to its possible injury, and to prevent the certainly increased difficulty of their removal under its weight. Sometimes even powerful winches and three ten-ton hydraulic jacks were ineffectual to remove the large stones from under the edges and the partitions. Boulders had to be undermined and rolled into the caisson, and there split by drills and wedges.

A vivid picture of the sensations and impressions within the caisson has been left by E. F. Farrington, the master mechanic on the construction of the bridge, a veteran of Roebling's bridges at Niagara and Cincinnati: "Inside the caisson," he wrote,

> everything wore an unreal, weird appearance. There was a confused sensation in the head, like "the rush of many waters." The pulse was at first accelerated, and then sometimes fell below the normal rate. The voice sounded faint and unnatural, and it became a great effort to speak. What with the flaming lights, the deep shadows, the confusing noise of hammers, drills, and chains, the half-naked forms flitting about, with here and there a Sisyphus rolling his stone, one might, if of a poetic temperament, get a realizing sense of Dante's inferno. One thing to me was noticeable—time passed quickly in the caisson.

The results of the first month's work were not very encouraging. The hardpan under the caisson was difficult to excavate, and the penetration attained was insufficient to seal the cutting edge securely. Frequent "blowouts" occurred —a large quantity of compressed air suddenly escaping under the edge. The body of compressed air was always trying to escape. Before the caisson became firmly fixed by side friction in the material penetrated, it was liable to lift up at one edge and thus set free a portion of the confined air. Sometimes the wave of a passing steamer would be sufficient to upset the equilibrium of the caisson and cause such a blowout. This sudden rush of escaping air would send up a large waterspout, thirty to sixty feet high, flooding a part of the work on top and stampeding the workmen laying the masonry. Inside the caisson the roaring was also heard, and with it came a powerful rush of air toward the place of escape. After each blowout the bottom of the working chamber would be flooded and the remaining air would be filled with a thick fog which persisted for some time.

The rate of descent under these handicaps—hardpan, boulders, and blowouts—averaged less than six inches a week. At this rate it would have taken more than a year and a half to sink the Brooklyn caisson, and about three years for the one on the New York side. And the wear and tear on tools was enormous.

But Colonel Roebling, who spent much time with the men in the caisson, was not disheartened. He realized that the initial rate of progress was disappointing. "On the other hand," he pointed out, "we are gaining daily in experience. The workmen are becoming more accustomed to the novel situation and more practised in the particular kind of work to be done."

There were new, increasing, and almost endless difficulties.

After the caisson reached the depth of 25 feet below the outside water level, the boulders encountered were so large and numerous that blasting was necessary. Although the idea of using powder had been entertained for some time, it had not yet been attempted as it appeared to involve too many hazards. There was no previous experience as to the possible consequences. What would be the effect of the explosions in the dense atmosphere? Would the concussions rupture the eardrums? Would the powder smoke in the confined space become suffocating? Would the blasts break the valves and doors in the air locks, letting the air escape and thus causing a flooding of the caisson? Would the pressure of the explosion cause a disastrous blowout of the water shafts? This last was the principal apprehension, as it would be fatal to both the men and the caisson.

Colonel Roebling courageously determined to settle these questions by actual personal experiment. He spent hours in different parts of the working chamber, firing off his revolver with successively heavier charges. Then he proceeded to fire small blasts of powder with a fuse, and finally he tried heavy charges of explosive. When the concussion in all cases was found to be harmless, blasting became an established procedure. The good effects were at once apparent in the expedited lowering of the caisson, twelve to eighteen inches a week instead of six inches. As many as twenty blasts were fired in one watch, the men merely stepping into an adjacent chamber to escape the flying fragments. One convenient way of disposing of boulders encountered under the shoe was to drill a hole through them, plant the explosive charge at the bottom, and shoot the rocks bodily into the caisson where they were broken up at leisure.

When blasting was used, great care had to be exercised against setting fire to the yellow-pine roof of the working chamber through the flash and the burning fuse. The gas pipe was broken several times, but the flame was extinguished before any real damage was done. In blasting under the shoe there was danger of injuring it, but nothing serious resulted. In fact, the shoe was already injured; armor plates were bent and crushed and partly torn off by jagged points of rock, the inner casting was cracked, and in many places the whole iron shoe had been forced in; yet no air escaped, because the clay was tight outside.

The work was carried on day and night by relays of men. The majority of the men took their meals along and remained down the full eight hours without any injury.

On deck there were double shifts of enginemen and firemen to run the excavating engines and the engines for the dirt cars, as well as two gangs to dump the cars. In addition there were the enginemen for the air compressors and the stone-hoisting engines, besides blacksmiths, machinists and gas men, one gang to remove the boulders brought up by the buckets, a carpenters' force of 25 men, and thirty men for setting the masonry. The total daily force at this tower foundation amounted in all to 360 men.

The work was so hard on the dredge buckets that the night shift was devoted to repairs, not only to the buckets but to the cars, engines, and other machinery as well. The men in the night gang also devoted themselves to getting out boulders from under the caisson in difficult places where only a few men could work, or they would run one dredging shaft alone, or be digging out the "well" under the other shaft.

As each air lock for entering the caisson held thirty men, two sets of lockings were required to let down a shift of 120 men. In changing shifts, the old gang remained in the working chamber below until relieved by the new shift.

Each of the two large water shafts came down through the roof of the caisson and extended about two feet below the level of the cutting edge into a "shaft well," or pool. By having the lower end of each shaft sealed in the well at the bottom, a balanced column of water was maintained in the shaft to serve as a seal against the escape of air from the working chamber. Into this well all excavated material in the caisson was dumped, and then removed to the top with a clamshell dredge or grab bucket operating through the shaft.

Gradually but surely the caisson sank toward its final resting place. At the end of five months 20,000 cubic yards of earth and stone had been removed.

Colonel Roebling was disappointed with the rate of progress. "In regular harbor dredging," he observed, "one bucket alone will raise 1000 yards per day. Our buckets therefore should have removed the material quite comfortably in one month's time. In place of one month five were required, and these were five months of incessant toil and worry, everlasting breaking down and repairing, and constant study where to improve if possible. We had in fact a material which could not be dredged."

The bucket could easily lift out any stone it could grapple, even up to one or two yards in size, provided the stone was shoved into the right position. Sometimes a large boulder would get stuck under the water shaft where the dredge could not reach it. Then someone would dive under the water and move the boulder—this for $2.00 a day and because of devotion to Roebling. "When the lungs are filled with compressed air, a person can remain under water from three to four minutes," said the Colonel. He knew—he had tried it.

The buckets were provided with heavy steel teeth. Different patterns of teeth were required for grappling stone, for scooping up mud, and for other materials. Constant repairs were necessary; a supply of five buckets was required to keep two in working order.

When the two halves of the bucket closed tight, it would bring up water and mud. But the smallest stone coming between the jaws would permit all the sand or other fine material to escape before the bucket reached the top. "And yet, with all our drawbacks," the Colonel remarked, "our daily experience confirmed us in the assurance that we had selected the only instrument capable of disposing of all the material at hand, no matter how large or ill-shaped or badly packed the boulder; no matter how tenacious the clay and hardpan, nor how flowing the occasional veins of quicksand."

At the beginning the air pressure was governed entirely by the tides, and regulated itself according to their height. During the falling of the tide it was practically unnecessary to run the air pumps; but with a rising tide the pumps had to be run at full speed to build up a balancing pressure. A declining pressure in the working chamber was always attended by a thick fog, which lasted until the tide changed. These fogs could not be entirely overcome, although pumping in a large excess of air partially dispersed them. Every change in the air pressure produced either a rise or fall of the column of water in the water shaft, as in a giant barometer; and constant attention was required to prevent a blow-out from the lowering of the water in the pool or from a washing away of the dam surrounding the pool.

When the caisson finally entered the watertight and airtight stratum of clay, the tides no longer had any effect upon the air pressure. The clay proved so tight that the pressure could easily be raised to four or five pounds higher than that called for by the outside head of water.

This satisfactory condition continued until fresh-water springs were encountered. Then there was an end to regularity of air pumping. Sometimes three pumps were sufficient, then again all six compressors running at their maximum speed could not maintain an adequate pressure. And as the pumps were overtaxed, they broke down, and the pressure fell still further. There was one critical period when the pressure fell to eleven pounds in place of twenty, where it should have stood.

A minimum of three pumps were needed at all times, in order to supply the necessary fresh air for 120 men and the numerous candles and gas lights, and to prevent a rise in temperature. The thermometer in the working chamber stood uniformly at 78 degrees, day and night, winter and summer, whether the temperature outside was ninety or zero.

With all the discomforts and exposure of working and blasting so far under water and the ever-present dread of being crushed under the great weight overhead or of being trapped in a flooded caisson, another grave danger arose from living in the compressed air—the danger of caisson disease, commonly called the "bends." The symptoms of an attack were cramps and severe pains in the joints, followed by paralysis. The caisson was no place for men suffering from heart or lung disease, or enfeebled by age or intemperance. But a man of good health and physique, with sound head, heart, and lungs, temperate in all things, and observing a few simple rules, had no difficulty in working under a pressure of twenty or thirty pounds.

In the depth reached by the Brooklyn caisson, the highest pressure ever attained in the air chamber was only 23 pounds to the square inch, added to the normal atmospheric pressure. Such a pressure was not sufficient to produce any serious cases of caisson disease. Only six men suffered attacks amounting to temporary paralysis, and in each case this occurred on their first visit and after staying below for only a short time. None of the old hands were affected to any extent, even when staying down eight hours.

The only physical discomfort normally experienced in the work on the Brooklyn caisson was in the effect on the eardrums in passing through the air lock when entering or leaving the caisson. During the period of sharply changing pressure in an air lock a severe pain in the ear is usually felt until the pressure on the two sides of the eardrum is equalized. To some the pain is insupportable, while others are hardly inconvenienced. But this is only a transient discomfort, although it may sometimes result in deafness.

As the weather became colder, the men became subject to colds and congestion of the lungs by reason of the severe drop in the air temperature from eighty to forty degrees which attended a passage out of the caisson through the air lock. A simple and effective remedy was provided by equipping the air lock with a heating coil, through which steam was allowed to flow as soon as the outlet air cocks were opened. This overcame the drop in temperature and, at the same time, eliminated the formation of the disagreeable mist which otherwise attended decompression.

As the caisson slowly descended, the tower masonry was being built up above by means of three large derricks with horizontal booms, standing on the caisson itself and guyed from the land. They controlled all parts of the foundation. For the lower courses Kingston limestone was used. Above low water granite was used exclusively for face stone. A large stone yard was established three miles below the bridge. As the stones arrived from the quarries in sailing vessels or barges, they were unloaded and assorted in courses, and then reloaded on stone scows and sent to the tower.

The load on top and the pressure within kept slowly increasing as the sinking progressed, until one Sunday morning, about 6 A.M., the "big blowout" occurred. Every particle of compressed air suddenly blew out through the south water shaft, leaving the caisson in an instant. The immediate cause was the washing away of the dam surrounding the pool of water under the shaft, and was the result of a watchman's negligence. As Colonel Roebling later pithily remarked: "To say that this occurrence was an accident would certainly be wrong, because not one accident in a hundred deserves the name. In this case it was the legitimate result of carelessness."

Luckily it occurred when no one was working below. What effects were produced inside the caisson at the time is therefore not known. Eyewitnesses outside heard an explosive roar and suddenly saw a dense column of water, fog, mud, and stones thrown up five hundred feet into the air, accompanied by a shower of falling fragments and yellow mud covering the houses for blocks around. This column was seen a mile off. The accompanying roar was deafening. The effect was like a volcanic explosion. The noise was so terrifying that the whole neighborhood was stampeded and made a rush up Fulton Street, to get away from the waterfront. Even the toll collectors at the ferry abandoned their tills and ran. The stampeding crowd soon collided with another, running in the opposite direction to see what had happened.

There were three men on top of the caisson at the time, including the watchman. The current of air rushing toward the blowing water shaft was so strong as to knock him down; while down, he was hit on the back by a stone, and then he lost consciousness. One of the other men jumped into the river, and the third buried himself in a coal pile.

It was all over in an instant. Both doors of the air lock automatically fell open wtih the release of interior pressure, and the dry bottom was visible through the air and water shafts. For the first and only time daylight entered the caisson and the interior could be seen without artificial illumination. Not a particle of water had entered under the shoe! As soon as possible a stream of water was passed into the water shafts from above, the locks were closed, and in the course of an hour the air pressure was restored to fifteen pounds. It required some courage to re-enter the caisson.

The total settling that had taken place during the few moments of this blowout amounted to ten inches. Every block under the frames and posts was absolutely crushed, the ground being too compact to yield; none of the frames, however, was injured or even out of line. The brunt of the blow had been taken by the shoe and the sides. As the loaded caisson suddenly descended, one sharp boulder under the cutting edge had cut the armor plate, crushed through the shoe casting, and buried itself a foot deep in the heavy oak sill. At this point, and in a number of other places, the sides of the caisson had been forced in some six inches. The marvel was that the airtightness was not impaired in the least.

Subsequent examination showed that the roof of the air chamber had assumed a permanent depression or sag of four or five inches. This deflection never increased.

As the caisson proceeded on its downward course, the disproportion between the dead weight above and the supporting air pressure below became greater and greater; and in order to meet this overweight, a large number of additional shores were introduced; each prop rested upon a block and wedges and supported a cap spiked against the roof. These shores, requiring repeated readjustment, added considerably to the labor of lowering the caisson; they also diminished the available working space. They gave, however, a positive assurance against any crushing of the caisson from above and could be readily removed when a boulder was to be taken out.

The downward movement of the caisson was usually so impulsive that the blocks under the posts were allowed to crush and were subsequently dug out. "In fact," remarked the Colonel, "their crushing was the only indication we had that any portion of the caisson was bearing particularly hard. The noise made by splitting of blocks and posts was rather ominous, and inclined to make the reflecting mind nervous in view of the impending mass of 30,000 tons overhead."

When the caisson had arrived within three feet of its proposed resting place, the Colonel decided to erect for its support 72 brick piers, symmetrically located. Their capacity was just sufficient to support the whole weight above in case the air should blow out. These pillars, blocking up the roof of the working chamber, were completed in three weeks. The wise foresight of this precautionary provision was later demonstrated.

In those days there were no electric lights—they were not invented by Mr. Edison until nine years later. But the men in the interior of the caisson could not work in darkness. The problem of satisfactorily illuminating the working chamber filled with compressed air presented many serious difficulties. In the absence of reflecting surfaces, a powerful light was required to penetrate the thick mist usually present and to illuminate every foot of the uneven soil.

Colonel Roebling tackled this problem with scientific practicality. Candles were tried first, but they were expensive and inefficient, and, from their rapid and incomplete combustion, they emitted an amount of smoke that was intolerable in the confined space. The inhaling of so much floating carbon was injurious, as the lampblack would remain in the lungs for weeks and months. Nevertheless, candles had to be used for all special work requiring close illumination. The Colonel found that he could eliminate much of the nuisance of the candle smoke by redesigning the candles and by special chemical treatment of the tallow and the wick.

Oil lamps he promptly discarded since they smoked more than candles, and the oil was dangerous as a fire hazard.

The existence of an establishment in New York for the production of compressed oxygen gas in large quantities and at moderate prices made the introduction of calcium lights quite feasible. A double system of pipes was put up in the air chamber, one for oxygen and the other for coal gas. Fourteen calcium lights were installed, two in each compartment. In addition there were sixty burners for common street gas, which was used whenever the supply of compressed oxygen failed. The total lighting cost was $5,000, of which the candles cost more than half.

The calcium lights had the advantage of producing less heat than gas or candles and practically no smoke, the product of combustion being principally water. The ordinary gas lights were far more economical than either calcium lights or candles, but they produced an intolerable amount of heat and vitiated the air even more than candles, although producing but little visible carbon.

To overcome the air pressure within, all gas had to be compressed before being fed through the pipes entering the caisson. Quite an apprenticeship was necessary to adapt the calcium lights to all new conditions. The chief danger lay in leaky pipes and in carelessly leaving cocks open. One gas explosion took place below, sufficient to singe off whiskers and create some alarm. The sense of smell became so blunted that leaking coal gas was not easily detected.

Naturally, one of the gravest dangers attendant upon caisson work before the days of the electric light was fire. The menace from fire, accelerated by the compressed air, was ever present. In the compressed air the flame of a candle would return when blown out. Every precaution had to be taken and all inflammable materials banished. Several minor fires at the beginning showed the necessity of constant caution. One of these early fires was of sufficient magnitude to demand the flooding of the caisson; this was easily accomplished at that time because the water entered freely under the shoe as the air escaped above through emergency release valves provided for the purpose.

But when the caisson entered more impervious material and the fact was settled that the river water was permanently excluded from the interior, it became a matter of increased importance to provide every possible safeguard against fires. Two large hose connections were installed, throwing high-pressure fire streams; and steam pipes were introduced, for smothering a blaze. In addition all seams between the roof timbers were carefully pointed with cement, and iron shields were placed over the permanent lights. It was made the special duty of two men to watch continually over all lights.

Notwithstanding all these precautions, there was a seam, where the supporting frames joined the roof, that had not been pointed; an empty candle box was nailed under it, in which some man kept his dinner, and while getting it he probably held a candle too close to the timber roof. This proved to be the heel of Achilles. The oakum calking ignited, and the air pressure drove the fire into the interior of the roof timbering and out of sight. The fire was discovered on the evening of Friday, December 2, 1870. Being directly over the frame, it had remained undetected until the latter had been partly burned through and, judging from the headway attained, it had been burning for some hours. A fire under the waves of the East River! A fire in timbering precariously supporting many thousands of tons of masonry overhead! A slight panic ensued among the men, but no one left the caisson.

Colonel Roebling was summoned, and he took charge of the situation. All appliances for putting out the fire were brought to bear. While the hose was being got ready, two large cylinders of carbonic-acid gas under high pressure were discharged into the burning area without producing any effect whatever; as soon as the stream of carbon dioxide was stopped the timber would immediately reignite. The two high-pressure streams of water were then turned on and these soon extinguished all fire that could be seen. There was a violent draft of air through the burned aperture; this was stopped with cement. The streams from the two hose connections were kept playing for two hours. At the end of that time one of them was replaced by steam at high pressure and allowed to run for half an hour; but it could not be ascertained if this was of any benefit, and the water was again turned on.

Colonel Roebling remained in the subaqueous chamber for seven hours, from ten at night to five in the morning, until the fire appeared to be conquered. Then he collapsed and was brought out unconscious. But four hours later he was back in the air chamber.

The question of flooding the caisson had been seriously discussed by the Chief Engineer and his assistants. To extinguish the fire without having recourse to flooding was very desirable, since that operation involved the risk of permanent injury to the caisson; on the other hand, if the fire was not out, it was simply a question of time before the entire structure would be destroyed.

The only way to ascertain the presence of fire was to bore for it at random through the solid timber. Colonel Roebling had therefore given instructions, before he collapsed, to have a series of auger holes bored upward into the timber roof.

A number of holes were bored for a distance of two feet. They showed no fire. Others were then bored three feet, still showing no fire. This result was, of course, encouraging. Time was lost in lengthening out augers and also in the boring, because the draft carried the chips up into the holes. At 8 A.M. a hole four feet deep revealed the dreaded fact that the fourth course of timber was one mass of living coals!

The Colonel, still weak and exhausted, reached the spot. He at once decided to flood the caisson as the only way to save it. The city fire department was called out and all available fire engines were soon at work pouring water down the shafts. Additional fire-fighting forces were brought up. The *Fuller,* a harbor fire boat, supplied eight powerful streams; the *J. L. Tebo* three, and a Navy Yard tug two more. By ten that morning 38 streams of water were pouring into the caisson, besides the water from the emergency pipes in the caisson itself. In five and a half hours the air chamber was filled. The total quantity of water required was over a million gallons.

The caisson was allowed to remain flooded for two and a half days. The air pressure was then again put on and the water forced out. It took six hours, and the air pressure required was twenty-two pounds. An inspection below showed but little apparent damage, beyond blocks that were crushed and some posts thrown over. The structure proved tighter than before the flooding, owing to the swelling of the timber. The building of the brick piers under the roof was resumed and completed in about two weeks, when the caisson was lowered down to them through the remaining distance of two feet.

No injury appeared in the masonry above from the changes in pressure to which the caisson had been subjected, but it took three months to make repairs below.

For several weeks after the fire, odors of turpentine and of other products of the combustion of yellow pine were very strong above the caisson, being forced out with the air bubbles; and a large quantity of frothy pyrolignic acid, a product of the destructive distillation of wood, made its appearance on top of the masonry. These signs gave rise to very unpleasant suspicions.

About two hundred exploratory borings were made in the roof to ascertain the extent of damage from the fire, both laterally and vertically. It was found that the damage was confined to the third and fourth courses of timber, but had spread out laterally in many directions, covering a much larger area than had been anticipated, the remotest points being some fifty feet apart. The fact that the air rushed out through every bore hole indicated that the immediate roof must be relied on to retain the compressed air. With this object in view it appeared advisable to inject cement grout into the burned cavities through the auger holes until the leakage was stopped. Accordingly, a cylinder was prepared with a piston and injecting pipe, and by this means 600 cubic feet of cement were forced into the cavities. The escape of air then ceased.

A number of trial bore holes failed to disclose any space not filled with hard cement; but in order to make sure, one large hole was cut up into the roof through five timber courses, directly over the place where the fire originated. It was found that the cement grout had, indeed, filled all vacant spaces, but that the timber was covered with a layer of soft and brittle charcoal varying from one to three inches in thickness. There was no alternative now but to remove all the cement and carefully scrape the charcoal from every burnt stick of timber. As the cement was removed, the full extent of the fire was gradually revealed, and in place of one opening in the timber, five were needed in order to reach the remotest points. This task required eighteen carpenters, working day and night for two months, in addition to the attendance of common labor.

Above the first opening the fire had destroyed the third, fourth, and fifth courses, having burned through the tin that had been placed between the fourth and fifth layers, and had also damaged parts of the sixth course. The fourth course, under the tin, had been the principal sufferer. The various ramifications of the fire had evidently been caused by air leaks. In several places one stick was burned away while the adjoining ones remained sound. The "fattest" sticks had succumbed the soonest. The general combustion had been of the nature of a slow charring, progressing in all directions.

The timbers were cleaned and squared off. Whenever a stick was found only partially consumed, it was carefully scraped and the cavity rammed full of cement. The larger spaces were filled up with lengths of yellow pine forced in with screw jacks and wedges, and well bolted vertically and laterally.

It was a job of dentistry on a mammoth scale—cleaning and filling a cavity worm-eaten by the erratic, zigzag wandering of the fire over a space of fifty feet square. The workmen had to crawl and work in all sorts of cramped areas and uncomfortable positions, lighted by little bull's-eye lanterns, in an air full of coal and cement dust and smoke. After everything was filled up solid, a number of long bolts were driven up from below so as to unite both the old and new timber into a compact body, and forty iron straps were bolted against the roof from below for additional reinforcement.

Under Colonel Roebling's zealous supervision the repair work was scrupulously and conscientiously executed. To set at rest any lingering doubts concerning the stability of the structure after the repairs, he called attention to the vast reserve of strength he had provided from the start. "It must be remembered," he observed, "that there are still eleven courses of sound timber above the burnt region. These have abundant capacity to distribute any local deficiency in equal bearing. From the faithful manner in which the work has been done, it is certain that the burnt region has now been made fully as strong, if not stronger than the rest of the caisson."

Two weeks after the "great fire" the excavation under the caisson had been completed, with the cutting edge at a depth of 44½ feet below mean high tide, and with the timber roof bearing on the brick supports. The concrete filling of the working chamber was then started, commencing at the shoe and proceeding in layers six to eight inches thick. Each layer was allowed to harden for five hours before the next was put on.

When the concreting had been in process a fortnight, one of the two supply shafts suddenly blew out. A charge of broken stone and gravel had jammed in the shaft; the men dumped in another load without measuring the depth, and then gave the signal to the man below, without shutting the upper door. The second charge loosened the first, and the two together overcame the air pressure against the lower door when the release lugs were turned. As soon as this happened, the air rushed out of the caisson, carrying an eruption of stone and gravel with it. Panic-stricken, the men above ran away, leaving those below to their fate. Anyone above, with the least presence of mind, could have closed the upper door by simply pulling at the rope.

Colonel Roebling was one of those trapped in the working chamber in this terrifying situation, between the timber roof and the rapidly rising water. His coolness and presence of mind in the emergency saved the day. He closed the lower door of the supply shaft, which had been jammed open, and soon everything was all right. But perhaps his cool command of a situation showed more vividly in this adventure than it ever had on the battlefield of the Civil

War. A vivid account of the experience is recorded in his own words:

> I happened to be in the caisson at the time. The noise was so deafening that no voice could be heard. The setting free of water vapor from the rarefying air produced a dark, impenetrable cloud of mist and extinguished all the lights. No man knew where he was going; all ran against pillars or posts or fell over each other in the darkness. The water rose to our knees and we supposed of course that the river had broken in. I was in a remote part of the caisson at the time; half a minute elapsed before I realized what was occurring and had groped my way to the supply shaft where the air was blowing out. Here I was joined by several men in scraping away the heaps of gravel and large stones lying under the shaft, which prevented the lower door from being closed. The size of this heap proved the fact of the double charge. From two to three minutes elapsed before we succeeded in closing the lower door. Of course everything was all over then, and the pressure, which had run down from seventeen to four pounds, was fully restored in the course of fifteen minutes. A clear and pure atmosphere accompanied it. The effect upon the human system, and upon the ears, was slight, no more than is experienced in passing out through the air lock.

It was fortunate that the interior brick piers had been provided and that the caisson had already been brought to a full bearing on them. Careful examination was made to see what effect the suddenly applied weight of 30,000 tons had had upon these pillars and the other supports. Although the brick piers had been subjected to an enormous concentration of pressure, they showed no signs of yielding. In the neighborhood of the water shafts and supply shafts, where there were no pillars or other supports, a slight local depression had taken place in the roof. The occurrence demonstrated the fact that the brick props were strong enough to bear the whole load, and also proved the necessity for their erection. There was no increase of air leakage from the abnormal strains to which the caisson had been subjected.

The Colonel, in his account of this experience, raised the interesting question: "What would have been the result if the water had entered the caisson as rapidly as the air escaped?" And he drew this significant conclusion:

"The experience here showed that the confusion, the darkness, and other obstacles were sufficient to prevent the majority of the men from making their escape by the air locks, no matter how ample the facilities." He pointed out, however, that there was an automatic safety provision: The supply shafts projected two feet below the roof into the air chamber; consequently, even if the caisson were flooded, one or two feet of air would be trapped in the top of the caisson so that "there would be enough air space left to save all hands who retained sufficient presence of mind."

The filling of the working chamber with concrete proceeded. A hundred cubic yards were laid per day, but as the space became smaller this rate of progress was reduced. A great saving in time as well as in concrete was effected by letting the edges of the caisson sink into the ground three feet deeper than the average level of the bottom. This diminished the volume to be filled about one-third.

During most of the time the weather outside was so cold that the concrete had to be mixed below. The gravel, being full of frost, frequently froze fast in the supply shaft; this was obviated by introducing a steam pipe and thawing out each charge. Later, when the weather became warmer, the concrete was mixed above and sent down directly for placing; much labor was thereby saved in the air chamber. The boulders which had been taken out of the caisson were broken up into square blocks, and again built in below with the concrete.

Next to the roof a shallow, sloping layer was rammed in with narrow, flat-faced iron rammers. As this was the most important point in all the filling and was apt to be slighted, it required the most careful watching.

The rapid influx from fresh-water springs prevented any material reduction of the air pressure. The water rising in the shafts was perfectly fresh, without a trace of salt, which showed the welcome fact that the timbers would ultimately be saturated with fresh water. Finally, when the working chamber had been filled in, the air locks were removed and the water shafts were filled with concrete.

When the sinking operation was started, the caisson had been loaded down with masonry, and additional courses were laid as the excavation in the working chamber proceeded. The building up of the tower masonry thus kept pace with the descent of the caisson, so that the work was always above water. When the caisson reached its final level, eleven courses of masonry had been laid. They were set in heavy beds of cement and all the spaces filled with either cement or concrete.

Colonel Roebling, scientifically trained, insisted on a careful planning and arrangement of engines, tracks, turntables, and the like for economical and efficient working. For these things he uniformly made a thorough study and worked out complete plans in advance. His engineering drawings, not only for the structure but also for the details of the construction equipment, were eagerly sought and published at the time in leading technical periodicals on both sides of the Atlantic. In his planning of the operations there was no fumbling, no cut and try, no guesswork. He handled the work like a problem of military supply. In consequence, the mechanics and laborers were never stopped for lack of materials, notwithstanding the difficulties of transportation and the limited space available for storage and handling.

He inherited his father's passion for safety. One rule which he followed throughout the building of the bridge was that no temporary work, such as derricks, scaffolding, and the like, should be used without a careful proportioning of the minutest detail to the strains involved. As a result it may truly be said that, throughout the thirteen years of construction, no man was injured by the breaking of any temporary work from insufficiency of design. Whatever accidents occurred resulted from invisible defects or from the apparently inevitable human equation.

With the boom derricks used for laying the masonry,

numerous guys were required, with hundreds of connections, any one of which giving way might result in a serious accident. Colonel Roebling established a rigorous system of daily inspection of all the equipment. Despite all precautions, however, a serious accident occurred in October, 1871. A defective weld in one of the connections gave way, causing two derricks to fall—killing three men and injuring five others.

During the intense cold of December plentiful use of hot water and salt kept the cement mortar from freezing and permitted the work to continue. By May, 1872, the masonry of the Brooklyn tower had been built up to a height of 75 feet above the river, with only twenty feet more required to bring the stonework up to the level of the bridge roadway.

The sinking of the caisson on the Brooklyn side had involved many trying experiences, but the foundation for the New York tower presented even greater problems, difficulties, and hazards. Layers of quicksand had to be conquered, a depth nearly twice as great had to be penetrated, and new and unknown terrors of caisson disease had to be faced.

The turning point for the Bridge—as determining the feasibility of the enterprise—was reached down in the earth and under the bed of the East River. Everything hinged on the success of this submarine operation, and the engineering world awaited the outcome.

The first borings John Roebling had made for the tower foundation on the New York side gave the appalling depth of 106 feet below the water level—a depth never before attempted. Finally a slight change in location reduced this by thirty feet.

The problem to be tackled was no simple one. Given a structural mass weighing from 30,000 to 50,000 tons, of steadily increasing height and weight, and covering an area of seven city lots, it was required, by burrowing underneath through a variety of uncertain materials, to lower the entire structure, including the superimposed tower masonry, steadily, accurately, and safely, to a firm foundation nearly eighty feet below the surface of the river.

The foundation on the New York side was located in deep water, at a distance of 400 feet from shore. It occupied the two slips of the Williamsburg Ferry. Nearly a year was consumed in negotiations for the removal of the ferry to an adjoining slip. Possession of the site was finally obtained in April, 1871, and the operation of dredging a level bottom for the reception of the caisson was immediately begun. The sloping river bed consisted of black dock mud above the sand, and was covered with sunken timber cribs filled with stone.

To explore the prospective foundation problems at this site, Colonel Roebling had made a series of borings down to rock, nine bore holes distributed around the area of 17,000 square feet. The conditions, as revealed by the borings, created a real foundation problem. Above the bedrock was a layer of cemented gravel of abundant resistance, interspersed with areas of flowing quicksand. And the rock itself was extremely sloping and irregular. To attempt to level it off by blasting under the caisson after that depth had been reached would be prohibitively costly in time, money, and hazard. It would add a year to the time of building the bridge and half a million dollars to the cost. To leave one corner of the foundation resting on the rock and the remaining area supported on more yielding material would invite disaster.

After studying the results of the test borings, the Colonel decided that the caisson must go nearly, if not quite, to rock. This determined the approximate depth to which the caisson would have to be sunk. The final question, whether to come to bearing on the rock or to stop a foot or more above it, would have to be settled when that point was reached. "The only course left open under the circumstances," he said, "is to proceed with the work and, when the caisson has arrived within a short distance of the rock, make a sufficient number of soundings, and then determine upon a course of action when we are face to face with the material."

As this caisson would ultimately be subjected to the load of a much larger mass of masonry than the one on the

Brooklyn side, he made the roof 22 feet thick instead of fifteen, to carry the greater weight. The working chamber was made of the same height as before, nine feet six inches.

To guard against fire—profiting by the experience of the Brooklyn caisson, in which seven fires had occurred, the last almost proving disastrous—the interior of the new air chamber was lined throughout with thin boiler iron, riveted together and calked. As a second safeguard a system of pipes was distributed through the structure in such a way that a large stream of water could be readily turned on at any point.

In this caisson the Colonel sought to improve his design of the air locks. He provided two sets of double air locks, enabling a whole working shift of 120 men to enter the chamber at one locking. And he placed the air locks at the bottom of the shafts, with the thought of saving the men the additional exposure of making the climb in compressed air. Moreover, in view of the greater height, a spiral stairway was installed instead of a ladder.

The plans for the New York caisson were perfected in the summer of 1870, and the contract for its construction was made in October. A year earlier, at the time of letting the contract for the Brooklyn caisson, it was with difficulty that any bids were secured. To launch a mass of those vast dimensions upside-down and broadside was so contrary to the experience of shipbuilders that they wanted large odds for the risk involved. But as a result of the increased confidence following the first successful experience, the second caisson, although larger and requiring much more work, was undertaken by the same builders for little more than half the cost of the first.

The caisson was built at the foot of Sixth Street, New York, the old yard across the river at Greenpoint having been abandoned for shipbuilding purposes. A severe winter, with delays in the delivery of the ironwork, prolonged the completion to May 8, 1871, on which day the launching took place. The caisson was then towed to the Atlantic

Basin, where additional courses of timber and concrete were put on preparatory to its removal to its permanent site.

Owing to the vexatious delays in obtaining possession of the ferry slips, nothing of any importance could be done toward locating machinery and workshops at the tower site until August, when the ferries finally stopped running. During the next few weeks a pile dock was built, as well as a large square enclosure of sheet piling to receive the caisson. On the pile platform, between the shore and the caisson berth, were erected engine houses, hoisting engines, unloading derricks, railroad tracks and hauling engines, concrete mixers, engines for pumping illuminating gas, thirteen air compressors for supplying air to the caisson, buildings for machinery and offices, sheds for blacksmiths, carpenters, and machinists, and for cement and other stores, and wash rooms, dressing rooms, a hospital, and rest bunks for the caisson men. All preparations for receiving the caisson were completed by September 11th, and on that day it was towed from the Atlantic Basin to its final resting place.

The compressed air produced peculiar effects upon sound. Pitch was raised. A deep bass voice became a treble, and the prolonged heavy sound of a blast was so modified as to resemble the sharp report of a pistol. A curious fact, noticed under these circumstances, was the impossibility of whistling: the utmost effort of the respiratory muscles could not increase the density of the air in the cavity of the mouth sufficiently to produce a musical note on its escape. A similar difficulty, though in a less degree, was experienced in speaking; and for this reason protracted conversation was very fatiguing.

For communication between the upper and lower world the telephone would have been of great benefit; but it had not yet been invented. Instead, an ingenious mechanical signal system contrived by the engineers proved of great assistance. It consisted of a tube with index pointers attached at top and bottom; under each pointer was a plan of the caisson, showing the position of every pipe and shaft; by rotating the tube, attention could be called to any one of these points in a moment. In addition, an index attached to a rod, both above and below, traversed a panel of signal messages, such as "Stop," "Start," "Bucket is caught," etc., directing what was to be done.

Above the caisson the stones were laid by three boom derricks, similar to those employed on the Brooklyn foundation. They were guyed solely from the caisson itself, so that the settling of the latter did not disturb the guys. For every twenty feet of masonry the derricks had to be raised, an operation requiring a few days.

The granite came from Maine and the limestone from Kingston, Lake Champlain, and Canajoharie. Owing to the early winter much of the stone in transit for the New York tower was frozen in; but this had been anticipated, and at no time was there any stoppage for want of stone.

During the severest weather the work of laying masonry was suspended for several days at a time. Down below, within the caisson, the excavation and sinking continued uninterrupted. Day in and out 236 men in two shifts were engaged here in removing the mud and gravel.

The performance of the dredge buckets in the mud and in the coarse sand and gravel was quite satisfactory. They constantly maintained a hole about six feet deep under each water shaft, and removed from three hundred to four hundred cubic yards of excavated material a day. The men were equipped with rubber boots, supplied by the bridge company, since this work required them to stand in the water.

Some 58 iron pipes to operate as "sand siphons" were located in the roof of the caisson, passing up through the timber and discharging above beyond the cofferdam. It was first intended to use water as the vehicle by which to eject the sand through these pipes; but, after a study of all the governing conditions, Colonel Roebling concluded that it would be more efficient to utilize the compressed air as the ejecting vehicle. He explained his reasoning to his staff and to the trustees:

> Another strong reason in favor of the air is this—An air

chamber with an iron skin can be made practically airtight, but a certain quantity of fresh air must be thrown in per minute to keep the atmosphere reasonably pure and fit to live in. This air would usually escape under the edges and do no work. Now, why not allow it to escape through pipes and at the same time carry out sand with it, and not be wasted? Any other mode of sending out the material would require extensive provision of additional machinery in the shape of pumps, boilers, and pipes, entailing an additional cost of at least $40,000, and would also involve difficulties of application for want of the required space around the foundation.

In view of all these considerations he decided to give the air-siphon system a thorough trial. The result vindicated his judgment, although new problems were created demanding the exercise of further inventive resourcefulness.

In the final arrangement a straight pipe extended to within one foot of the bottom of the caisson. The sand or gravel was heaped around the lower end of the pipe until it was two or three feet deep, when the cock at the top was opened, letting the rush of escaping air carry the material out. The regular supply of air was sufficient to keep three sand siphons going at a time, and the work of throwing and feeding the sand to these pipes was enough to keep sixty or more men busy. The labor in itself was very fatiguing, making frequent resting spells necessary.

When two or more stones jammed in the pipe, they were loosened by dropping a bar attached to a rope from above. When the pressure reached 28 pounds, it was found that too much air escaped, and the lower ends of the pipes were reduced to 2½ inches.

The material carried by the blast of air passed out of the vertical discharge pipe with tremendous velocity, stones and gravel often being projected at least 400 feet high. As the sand was very sharp, the constant abrasion soon caused the pipes to wear through near the roof.

To deflect the sand at the top of the pipes at right angles, both wrought- and cast-iron elbows were at first used. The furious blast of sharp sand would generally cut through these in an hour or two, sometimes in a few minutes, although the iron was 1½ inches thick. The plan finally adopted was to use an elbow into which a back piece of hard, chilled iron, about two inches thick, was fitted and secured. This device was cheap, and solved the problem. Under the action of the sand blast it would last as long as two days, and then it required but a few minutes for a man to remove the holding key and insert a new back piece to receive the blast.

Great care was required to prevent accidents from the use of the sand ejectors, as, with a heavy pressure, small pebbles were discharged almost with the velocity of musket balls. One of the men, by incautious exposure, was struck in the arm by a large fragment, producing a wound more than five inches long. A boatman on the river had his finger shot off.

Toward the end of the work, when the excavated material from below was utilized above as back fill around the masonry, the horizontal elbows of the sand siphons were all removed, and heavy granite blocks were set above the mouths of the pipes. The sand was then discharged against these granite blocks and thus deflected into the cofferdam.

When the quicksand was fairly entered upon, it was found that the dredge buckets in the water shafts became useless. The quicksand, in combination with small stones and boulders, would compact to a mass as hard as rock. This mass could not be penetrated by the teeth of a grab bucket, and even the point of a crowbar could scarcely be driven into it. A trial remedy was the use of a hose under the shaft to loosen and stir up the material, but then the sand was so fine and quick-running as to escape through the slightest crevice in the buckets. The sand pipes became thenceforth the sole reliance, until the coarse gravel and stone became so plentiful as to choke the ends of the pipes, making it necessary to stop to remove the stones. On this account the work during the last ten feet was very slow and tedious. Before this the progress had at times averaged a foot a day, but toward the end this rate decreased to one or two feet a week.

At a depth of about sixty feet a bone of the domestic sheep was found under a large boulder, and still lower down some cannon balls and fragments of brick and pottery. For an enduring foundation it was necessary to go deeper, in order to reach strata that had not been disturbed in the geologic age of man.

At 68 feet a number of boulders too large to be moved were encountered under one water shaft. To permit the sinking of the caisson to continue, the shaft was cut off near the roof of the air chamber, after blowing out the water and capping the top of the shaft. The same process soon had to be repeated with the other water shaft.

The downward movement of the caisson remained under perfect control throughout the whole of the sinking. It was very gradual, owing to the large bearing area presented by the wide cross frames and broad shoe which the Colonel had provided in his design for this caisson. While the caisson was passing through the mud, river sand, and gravel, the frames sank through the material without digging; but in the quicksand and harder material below, whole areas of the frames had to be dug out underneath before settlement would take place. The caisson was also controlled to sink perpendicularly in its true place, any side or tilting movement being promptly corrected by digging under the high side. Toward the end of the excavation fully half of the caisson rested on quicksand and the other half on cemented gravel; but even under these extreme circumstances the sandhogs, under Roebling's direction, were able to put the hard side down lower, while the soft side was kept stationary.

When a depth of 70 feet was reached, soundings were begun in the air chamber for the location of bedrock, by means of a pointed rod ten feet long, driven in by sledges. These thorough probings were carried on daily for a month, until a clear idea of the form and depth of the bedrock was attained. The surface of the underlying rock was found to be very irregular, composed of alternate projections and depressions, the extreme difference in elevations being sixteen feet. Fortunately the top of the rock was found to be

covered by a layer of very compact material—cemented
gravel—so hard that it was impossible to drive in an iron
rod without battering it to pieces. This was the first stratum
reached that could sustain the weight of the tower and
that bore evidences of not having been disturbed. It was
good enough to found upon, as good as any concrete that
could be put in place of it. Colonel Roebling therefore decided to stop excavation at a depth of 78 feet and to
rest the caisson on the hard material overlying the rock.
The jagged points of rock were leveled off and all material
was removed down to that depth.

The excavation and sinking were completed in May, 1872.
Everything was kept under perfect control. The caisson was
stopped exactly at the predetermined depth. The difference
of level at the extreme corners of the cofferdam on top of
the caisson was measured and found to be less than one
inch.

The concrete filling was done as in the Brooklyn caisson, except that the greater supporting power of the thicker
roof and the wider cross frames rendered the brick piers
superfluous, and they were dispensed with. In filling the
working chamber, the spaces under the frames were first
made solid; then, wherever there was quicksand under the
shoe, a trench was cut down to the hard gravel below and
filled with concrete. All the remaining quicksand was dug
out below, down to the hardpan, and worked into the concrete. The filling procedure was carried out in such a manner that the final exit could be made by means of the water
shafts.

For seven months, through winter and spring, Colonel
Roebling and his men had been working in the compressed
air of the submarine chamber. As the pressure increased,
the working hours below for each man had been gradually
reduced from the initial eight hours until finally, when the
pressure was 35 pounds, the men worked only two hours
at a time, twice a day.

Men were now hard to get, wages were high, and the time
of labor became so short that the progress of the work was

slow and costly. To make any real headway required an immense amount of supervision. The young bridgebuilder fully appreciated the problem on his hands. "The labor question," he remarked, "becomes the most serious drawback. The forces of nature may be measured and brought under control provided they are properly understood, but human nature is not so amenable to laws." Early in his training he had learned that the management of men is an important part of an engineer's work.

Ascending the circular stairway above the air lock was found to be exceedingly fatiguing as the depth increased, and the stairway in one of the shafts was replaced by a steam elevator. This relieved the men of the exertion of climbing at a time when their circulation was embarrassed and their system unstrung by the sudden return to normal pressure.

The men in the caisson suffered serious discomfort from the large amount of unconsumed carbon which floated in the air in the form of smoke from the burning gas jets. The inhalation of this suspended carbon produced irritation of the air passages and gave rise to a characteristic black expectoration. Six months after the work in the caisson was finished, some of the men were still coughing up sputa streaked with black.

By increasing the number of compressors, the atmosphere in the caisson was brought to a degree of purity such that it contained only one-third of one percent of carbon dioxide. To maintain this standard, 150,000 cubic feet of fresh air had to be pumped into the working chamber per hour.

The first observed physiological effects of the compressed air arose from the sudden unbalanced pressure on the eardrums. "Hold your noses," was the warning cry of the attendant before closing the door of the lock. The men were strenuously warned not to go into the lock unless they were able, when holding the nose and blowing forcibly, to feel air enter both ears through the Eustachian tubes. When this precaution was neglected, the individual was said to be "caught in the lock" or unable to "change his ears." Inability to equalize the pressure upon the two sides of the eardrum caused extreme pain, or acute inflammation, or sometimes a ruptured eardrum.

But the affliction to be dreaded was the painful paralysis, and sometimes death, that might follow upon a stay in the compressed air. In those days caisson disease was a new experience, and the medical profession was mystified. It was not until six years later that the nature of caisson disease was determined—by Paul Bert, a French physiologist. The paralysis was found to be caused by the liberation of nitrogen bubbles from solution in the blood and tissues of the body upon reduction of pressure. The danger lies, not in the compression within the air chamber, but in the decompression upon leaving the caisson. The rate of reduction of pressure is the critical factor; and in the case of an attack, the indicated remedy is prompt return to compressed air followed by slow decompression. In modern caisson or tunnel work "hospital locks" are provided at the site for this purpose.

In 1872, however, the cause and the treatment were still a mystery. A few facts were known: The onset usually came immediately upon emerging. Persons in a nervous and excited state, and those of a weak constitution, were most susceptible. While in compressed air, violent exertion such as climbing of ladders and other strenuous work must be minimized. At higher pressures the working period must be reduced.

When the higher air pressures were reached in the New York tower foundation, special precautions were taken against caisson disease; and medical attention was provided. Colonel Roebling engaged the services of Dr. Andrew H. Smith to attend to all caisson cases and to examine new candidates for work below.

Not until the pressure reached 24 pounds were any serious effects observed. Dr. Smith was in daily attendance. Every man on the job was examined, in order to exclude those suffering from weak hearts or lungs and those whose resistance was reduced by age or intemperance. All new men were required to secure a permit from the doctor before they were allowed to enter the caisson.

Strict "Rules of Health" were drawn up and were posted

conspicuously for the guidance of the workmen in the caisson:

1] Never enter the caisson with an empty stomach.
2] Use, as far as possible, a meat diet and take warm coffee freely.
3] Always put on extra clothing on coming out, and avoid exposure to cold.
4] Exercise as little as may be during the first hour after coming out, and lie down if possible.
5] Use intoxicating liquors sparingly; better none at all. It is dangerous to enter the caisson after drinking intoxicating liquor.
6] Take at least eight hours of sleep every night.
7] See that the bowels are open every day.
8] Never enter the caisson if at all sick.
9] Report at once to the office all cases of illness, even if they occur after returning home.

The physical conditions to which the men were subjected were a strain on their constitutions and on their powers of adjustment. In passing through the lock going down, they endured a sudden rise in temperature of fifty or sixty degrees, coinciding with an increase of atmospheric pressure of from 18 to 36 pounds, at the same time passing from ordinary air to air supersaturated with moisture. Within the caisson the breathing became deeper and faster. The work of the heart was increased and its action was therefore excited. The pulse rose to 120 immediately upon entering the caisson. After two to four hours of work in the compressed air the men had again to pass through the air locks, where this time they underwent a rapid reduction of pressure and temperature. The sudden chilling was alleviated by surrounding the interior of the lock with coils of pipe heated with steam, turned on only when locking out.

Hot coffee was always served to each man immediately he emerged from the caisson. It served as a warming stimulant to relieve the chill and the nervous prostration which marked the return to the open air.

A large room in the yard was fitted up with bunks to give the men an opportunity to rest while the changes of readjustment to normal were going on in the system.

Despite all the precautions taken, a large proportion of the workers suffered from the effects of the compressed air. From January through May there were 110 cases of sufficient severity to require medical treatment; and in three of these cases, the victims died.

The sick were treated in the first instance at the yard, where a room was set apart as a temporary hospital. Remedies were kept on hand: ergot, morphine, and stimulants. Cases occurring during the doctor's absence were treated by the engineer on duty, according to a prescribed plan; or, if the case was severe, the doctor was summoned. Serious cases were sent to the public hospitals.

The ages of the sandhogs ranged from 18 to 45, and they were of almost all nationalities. The habits of many of the men were doubtless not favorable to health, but everything which admonition could do was done to restrain them from excesses. Many of them slept in crowded, vermin-infested lodginghouses, where the beds or bunks were arranged in tiers, one above the other, in rooms in which there was scarcely an attempt at ventilation. One of the men fell a victim to spotted fever, contracted from such surroundings; and his death was at once ascribed by his comrades to the effect of the compressed air.

The onset of the bends invariably developed after the air pressure was removed. The attacks were characterized by painful cramps in one or more of the extremities, and sometimes in the trunk, frequently accompanied by headache and vertigo, epigastric pain and vomiting. The access of the first sharp neuralgic pains was often very abrupt, as if the patient "had been struck by a bullet." The pain was excruciating—men of the strongest nerve were completely subdued by it. It felt "as if the flesh were being torn from the bones." The agonizing torture generally began in the knees, shifting to the legs or thighs, then creeping up the trunk to seize upon the shoulders and arms. The skin was of a leaden hue, and was covered with a profuse perspiration standing out in beads upon the surface. In 15 percent of the cases, paralysis occurred. It most frequently affected the lower half of the body, but in some instances also included the trunk and one or both arms. Accompanying symptoms were headache, dizziness, double vision, incoherence of speech, and sometimes unconsciousness. The duration of the attack usually varied from a few hours to six or eight days, but in some cases it continued for weeks.

In the fatal cases a condition of profound coma was the usual forerunner of death. The occurrence of this symptom left but little hope of recovery.

The first fatal case occurred on April 22nd, when the caisson was at a depth of 75 feet. John Myers, about forty years old, a native of Germany, went to work in the caisson for the first time that day, the pressure being about 34 pounds. He was of a stout, heavy build, of "very full habit." When examined two days before, his lungs were sound. He worked through the morning shift of two and a half hours without inconvenience, and remained about the yard for half an hour after coming up. He then complained of not feeling well, and went to his boarding place, which was but a few rods distant. As he passed through the lower story of the house, on his way to his own room, which was on the second floor, he complained of pain in the abdomen. When nearly at the top of the stairs, he sank down insensible; and before he could be laid on his bed, he was dead.

Eight days later a second fatal case occurred. Patrick McKay, aged fifty, born in Ireland, had been working four months in the caisson and had not complained of ill health. On April 30th he remained in the caisson half an hour beyond the usual time, at the second watch, under a pressure of 34 pounds. The others who were with him in the air lock found, when about to emerge, that he was sitting with his back against the iron wall of the lock, insensible. He was at once carried up to the surface and removed to the Park Hospital. He was unconscious, with face pale and dusky, lips blue, pulse irregular and feeble. Under the administration of stimulants he recovered some degree of consciousness, and begged incessantly for water. Paroxysms of convulsions soon set in, and in one of these he died, nine hours after the attack. The autopsy showed that all of the organs were healthy, except the kidneys, which were the seat of Bright's disease and were very much altered in structure. "In this case," said the physician, "the effect of the compressed air was merely to hasten an event which, at best, could not have been long delayed."

The following month the third death occurred. William Reardon, born in England, aged 38, corpulent and "of intemperate habits," began work on the morning of May 17th. He was advised to work only one watch—two and a half hours—on the first day; nevertheless, feeling perfectly well after the first shift, he went down again in the afternoon. The pressure at this time was 35 pounds. Immediately after coming up from the second watch—two hours—he was taken with severe pain in the stomach, followed by vomiting. In a few minutes the cramps seized his legs, which soon lost the power of motion though they continued to be the seat of extreme pain. He was removed to the Centre Street Hospital, where he gradually sank, and on the following morning he died.

These fatalities from the work in the pneumatic foundation aroused excited criticism in the public press. The type of operation was new, caisson disease was a mystery, and there was no yardstick of prior experience. The public jumped to the conclusion that criminal negligence was involved.

Colonel Roebling did his best to present a clear picture of the situation, in the light of the limited medical knowledge of the subject then available. "The few cases of death that have occurred can in but two instances be charged to the effects of the pressure," he pointed out. "It is true that scarcely any man escaped without being somewhat affected by intense pain in his limbs or bones, or by a temporary paralysis of arms and legs; but they all got over it, either by suffering for a few days outside, or by applying the heroic mode of returning into the caisson at once, as soon as the pains manifested themselves, and then coming out very slowly."

In his official report on the progress of the work at this time, Washington Roebling paid generous acknowledgment to his assistants and to the workmen, adding these words pregnant with future significance: "The labor below is always attended with a certain amount of risk to life and health, and those who face it daily are, therefore, deserving of more than ordinary credit."

During the days and nights that the work was going on under the bed of the East River, the young chief engineer was continually in the caisson, personally directing the efforts of his men and setting an example of devotion and courage. From him flowed the energy and impulse that animated the entire force. Mindful always that any slip, no matter how trivial, at this stage of the work, might prove disastrous, he actually spent more hours in the working chamber than anyone else. Night and day he worked with the men under the crushing air pressure, until it wore out his strength.

One afternoon in the early summer of 1872 Colonel Roebling had to be carried out of the caisson, nearly insensible, a victim of the dread caisson disease. All night his death was hourly expected.

In a few days he rallied and, mustering all his reserve of strength, he attempted to return to work. But he again collapsed and had to be brought home.

At the age of 35 his days of physical exertion were ended. He remained painfully paralyzed—doomed to lifelong suffering—with progressive blindness and deafness setting in, vocal cords affected, and every nerve and muscle tortured with pain.

Thenceforth this man, who had been full of life and hope and daring at the inception of the work, was a crippled invalid, confined to his home. Just as his brilliance as an engineer was at its height, and just as his masterwork was passing its crisis, he was struck down.

The diversity of engineering activity is illustrated in the contrast between the lives of Grenville Dodge and Nikola Tesla. Tesla's adventures were primarily of the mind. He was adamant in his determination to put his invention of alternating current to the use of industry and mankind, but was utterly naïve in dealing with the practical aspects of his negotiations. Having first sold his patent rights to George Westinghouse, he later sacrificed millions of dollars in royalties by renouncing them out of motives of sympathy for Westinghouse. In later life, due perhaps to mental instability, he became what might be termed a technical charlatan, whose experiments were at variance with acceptable engineering practices. John J. O'Neill, author of Prodigal Genius, the Life of Nikola Tesla from which this selection is taken, was formerly Science Editor of The New York Herald Tribune and received the Pulitzer Award in Journalism.

TESLA AND ALTERNATING CURRENT

JOHN J. O'NEILL

TESLA, in 1875, at the age of 19, went to Gratz, in Austria, to study electrical engineering at the Polytechnic Institute. He intended henceforth to devote all his energies to mastering that strange, almost occult force, electricity, and to harness it for human welfare.

He completely eliminated recreation and plunged into his studies with such enthusiastic devotion that he allowed himself only four hours' rest, not all of which he spent in slumber. He would go to bed at eleven o'clock and read himself to sleep. He was up again in the small hours of the morning, tackling his studies.

Under such a schedule he was able to pass, at the end of the first term, his examinations in nine subjects—nearly twice as many as were required. His diligence greatly impressed the members of the faculty. The dean of the technical faculty wrote to Tesla's father, "Your son is a star of first rank." The strain, however, was affecting his health. He desired to make a spectacular showing to demonstrate to his father in a practical way his appreciation of the permission he gave to study engineering. When he returned to his home at the end of the school term with the highest marks that could be awarded in all the subjects passed, he expected to be joyfully received by his father and praised for his good work. Instead, his parent showed only the slightest enthusiasm for his accomplishment but a great deal of interest in his health, and criticized Nikola for endangering it.

Early in his second year at the Institute there were received from Paris a piece of electrical equipment, a Gramme machine, that could be used as either a dynamo or motor. If turned by mechanical power it would generate electricity, and if supplied with electricity it would operate as a motor and produce mechanical power. It was a direct-current machine.

When Professor Poeschl demonstrated the machine, Tesla was greatly impressed by its performance except in one respect—a great deal of sparking took place at the commutator. Tesla stated his objections to this defect.

"It is inherent in the nature of the machine," replied Professor Poeschl. "It may be reduced to a great extent, but as long as we use commutators it will always be present to some degree. As long as electricity flows in one direction, and as long as a magnet has two poles each of which acts oppositely on the current, we will have to use a commutator to change, at the right moment, the direction of the current in the rotating armature."

"That is obvious," Tesla countered. "The machine is limited by the current used. I am suggesting that we get rid of the commutator entirely by using alternating current."

Long before the machine was received, Tesla had studied the theory of the dynamo and motor, and he was convinced that the whole system could be simplified in some way. The solution of the problem, however, evaded his grasp, nor was he at all sure the problem could be solved—until Prof.

Poeschl gave his demonstration. The assurance then came to him like a commanding flash.

The first sources of current were batteries which produced a small steady flow. When man sought to produce electricity from mechanical power, he sought to make the same kind the batteries produced: a steady flow in one direction. The kind of current a dynamo would produce when coils of wire were whirled in a magnetic field was not this kind of current—it flowed first in one direction and then in the other. The commutator was invented as a clever device for circumventing this seeming handicap of artificial electricity and making the current come out in a one-directional flow.

The flash that came to Tesla was to let the current come out of the dynamo with its alternating directions of flow, thus eliminating the commutator, and feed this kind of current to the motors, thus eliminating the need in them for commutators. What he did not expect was to draw a storm of criticism.

Professor Poeschl, however, deviated from his set program of lectures and devoted the next one to Tesla's objections. With methodical thoroughness he picked Tesla's proposal apart and, disposing of one point after another, demonstrated its impractical nature so convincingly that he silenced even Tesla. He ended his lecture with the statement: "Mr. Tesla will accomplish great things, but he certainly never will do this. It would be equivalent to converting a steady pulling force like gravity into rotary effort. It is a perpetual motion scheme, an impossible idea."

Deep down in his innermost being, however, Tesla held firmly to the conviction that his idea was a correct one. Criticism only temporarily submerged it, and soon it came bobbing back to the surface of his thinking. He gradually convinced himself that, contrary to his usual procedure, Professor Poeschl had in this case demonstrated merely that he did not know how to accomplish a given result, a deficiency which he shared with everyone else in the world, and therefore could not speak with authority on this subject. And, in addition, Tesla reasoned, the closing remark with which Professor Poeschl believed he had clinched his argument—"It would be equivalent to converting a steady pulling force like gravity into a rotary effort"—was contradicted by Nature, for was not the steady pulling force of gravity making the moon revolve around the earth and the earth revolve around the sun?

"I could not demonstrate my belief at that time," said Tesla, "but it came to me through what I might call instinct, for lack of a better name. But instinct is something which transcends knowledge. We undoubtedly have in our brains some finer fibers which enable us to perceive truths which we could not attain through logical deductions, and which it would be futile to attempt to achieve through any wilful effort of thinking."

In his mind he constructed one machine after another, and as he visioned them before him he could trace out with his finger the various circuits through armature and field coils, and follow the course of the rapidly changing currents. But in no case did he produce the desired rotation.

A radical change had taken place in Tesla's mode of life while at Gratz. The first year he had acted like an intellectual glutton, overloading his mind and nearly wrecking his health in the process. In the second year he allowed more time for digesting the mental food of which he was partaking, and permitted himself more recreation. About this time Tesla took to card playing as a means of relaxation. His keen mental processes and highly developed powers of deduction enabled him to win more frequently than he lost. He never retained the money he won but returned it to the losers at the end of the game.

Instead of going to the University of Prague in the fall of 1878 as he had planned, Tesla accepted a lucrative position that was offered him in a technical establishment at Maribor, near Gratz. He was paid sixty florins a month and a separate bonus for the completed work, a very generous compensation compared with the prevailing wages. During this year Tesla lived very modestly and saved his earnings.

The money he had saved at Maribor enabled him to pay his way through a year at the University of Prague, where he extended his studies in mathematics and physics. He continued experimenting with the one big challenging alternating-current idea that was occupying his mind. He had explored, unsuccessfully, a large number of methods and, though his failures gave support to Professor Poeschl's contention that he would never succeed, he was unwilling to give up his theory.

It would have been a pleasure to Tesla to have continued his studies, but it now was necessary for him to make his own living. His father's death, following Tesla's graduation from the University at Prague, made it necessary for him to be self-supporting. Now he needed a job. Europe was extending an enthusiastic reception to Alexander Graham Bell's new American invention, the telephone, and Tesla heard that a central station was to be installed in Budapest. The head of the enterprise was a friend of the family. The situation seemed a promising one.

Without waiting to ascertain the situation in Budapest, Tesla, full of youthful hope and the self-assurance which is typical of the untried graduate, traveled to that city, expecting to walk into an engineering position in the new telephone project. He quickly discovered, on his arrival, that there was no position open; nor could one be created for him, as the project was still in the discussion stage.

It was, however, urgently necessary, for financial reasons, that he secure immediately a job of some kind. The best he could obtain was a much more modest one than he had anticipated. The salary was so microscopically small he would never name the amount, but it was sufficient to enable him to avoid starvation. He was employed as draftsman by the Hungarian Government in its Central Telegraph Office, which included the newly developing telephone in its jurisdiction.

It was not long before Tesla's outstanding ability attracted the attention of the Inspector in Chief. Soon he was transferred to a more responsible position in which he was engaged in designing and in making calculations and estimates in connection with new telephone installations. When the new telephone exchange was finally started in Budapest in 1881, he was placed in charge of it.

Tesla was very happy in his new position. At the age of 25 he was in full charge of an engineering enterprise. His inventive faculty was fully occupied and he made many improvements in telephone central-station apparatus. Here he made his first invention, then called a telephone repeater, or amplifier, but which today would be more descriptively called a loudspeaker—an ancestor of the sound producer now so common in the home radio set. This invention was never patented and was never publicly described, but, Tesla later declared, in its originality, design, performance, and ingenuity it would make a creditable showing alongside his better known creations that followed. His chief interest, however, was still the alternating-current motor problem whose solution continued to elude him.

Always an indefatigable worker, always using up his available energy with the greatest number of activities he could crowd into a day, always rebelling because the days had too few hours in them and the hours too few minutes, and the seconds that composed them were of too short duration, and always holding himself down to a five-hour period of rest with only two hours of that devoted to sleep, he continually used up his vital reserves and eventually had to balance accounts with Nature. He was forced finally to discontinue work.

The peculiar malady that now affected him was never diagnosed by the doctors who attended him. It was, however, an experience that nearly cost him his life. To doctors he appeared to be at death's door. The strange manifestations he exhibited attracted the attention of a renowned physician, who declared medical science could do nothing to aid him. One of the symptoms of the illness was an acute sensitivity of all of the sense-organs. His senses had always been extremely keen, but this sensitivity was now so tremendously exaggerated that the effects were a form of torture. The ticking of a watch three rooms away sounded like the beat of hammers on an anvil. The vibration of ordinary city traffic, when transmitted through a chair or bench, pounded through his body. His pulse, he said, would vary from a few feeble throbs per minute to more than one hundred and fifty.

Once the crisis was past and the symptoms diminished, improvement came rapidly and with it the old urge to tackle problems.

When the acute sensitivity reduced to normal, permitting him to resume work, he took a walk in the city park of Budapest with a former classmate, named Szigeti, one late afternoon in February, 1882. While a glorious sunset overspread the sky with a flamboyant splash of throbbing colors, Tesla engaged in one of his favorite hobbies—reciting poetry.

The prismatic panorama which the sinking sun was painting in the sky reminded him of some of Goethe's beautiful lines:

> The glow retreats, don' is the day of toil;
> It yonder hastes, new fields of life exploring;
> Ah, that no wing can lift me from the soil,
> Upon its track to follow, follow soaring. . . .

Tesla, tall, lean, and gaunt, but with a fire in his eye that matched the flaming clouds of the heavens, waved his arms in the air and swayed his body as he voiced the undulating lines. He faced the color drama of the sky as if addressing the red-glowing orb as it flung its amorphous masses of hue, tint, and chrome across the domed vault of heaven.

Suddenly the animated figure of Tesla snapped into a rigid pose as if he had fallen into a trance. Szigeti spoke to him but got no answer. Again his words were ignored. The friend was about to seize the towering motionless figure and shake him into consciousness when instead Tesla spoke.

"Watch me!" said Tesla, blurting out the words like a child bubbling over with emotion: "Watch me reverse it." He was still gazing into the sun as if that incandescent ball had thrown him into a hypnotic trance.

Szigeti recalled the image from Goethe that Tesla had been reciting: "The glow retreats . . . It yonder hastes, new fields of life exploring," a poetic description of the setting sun, and then his next words—"Watch me! Watch me reverse it."

Did Tesla mean the sun? Did he mean that he could arrest the motion of the sun about to sink below the horizon, reverse its action and start it rising again toward the zenith?

"Let us sit and rest for a while," said Szigeti. He turned him toward a bench, but Tesla was not to be moved.

"Don't you see it?" expostulated the excited Tesla. "See how smoothly it is running? Now I throw this switch—and I reverse it. See! It goes just as smoothly in the opposite direction. Watch! I stop it. I start it. There is no sparking. There is nothing on it to spark."

"But I see nothing," said Szigeti. "The sun is not sparking. Are you ill?"

"You do not understand," beamed the still excited Tesla, turning as if to bestow a benediction on his companion. "It is my alternating-current motor I am talking about. I have solved the problem. Can't you see it right here in front of me, running almost silently? It is the rotating magnetic field that does it. See how the magnetic field rotates and drags the armature around with it? Isn't it beautiful? Isn't it sublime? Isn't it simple? I have solved the problem. Now I can die happy. But I must live, I must return to work and build the motor so I can give it to the world. No more will men be slaves to hard tasks. My motor will set them free, it will do the work of the world."

Tesla was now a little more composed, but he was floating on air in a frenzy of almost religious ecstasy. He had been breathing deeply in his excitement, and the overventilation of his lungs had produced a state of exhilaration.

Picking up a twig, he used it as a scribe to draw a diagram on the dusty surface of the dirt walk. As he explained the technical principles of his discovery, his friend quickly grasped the beauty of his conception, and far into the night they remained together discussing its possibilities.

Alternating-current motors had heretofore presented what seemed an insoluble problem because the magnetic field produced by alternating currents changed as rapidly as the current. Instead of producing a turning force they churned up useless vibration.

Up to this time everyone who tried to make an alternating-current motor used a single circuit, just as was in direct current. As a result the projected motor proved to be like a single-cylinder steam engine, stalled at dead center, at the top or bottom of the stroke.

What Tesla did was to use two circuits, each one carrying the same frequency of alternating current, but in which the current waves were out of step with each other. This was equivalent to adding to an engine a second cylinder. The pistons in the two cylinders were connected to the shaft so that their cranks were at an angle to each other which caused them to reach the top or bottom of the stroke at different times. The two could never be on dead center at the same time. If one were on dead center, the other would be off and ready to start the engine turning with a power stroke.

This analogy oversimplifies the situation, of course, for Tesla's discovery was much more far-reaching and fundamental. What Tesla had discovered was a means of creating a rotating magnetic field, a magnetic whirlwind in space which possessed fantastically new and intriguing properties. It was an utterly new conception. In direct-current motors a fixed magnetic field was tricked by mechanical means into producing rotation in an armature by connecting successively through a commutator each of a series of coils arranged around the circumference of a cylindrical armature. Tesla produced a field of force which rotated in space at high speed and was able to lock tightly into its embrace an armature which required no electrical connections. The rotating field possessed the property of transferring wireless through space, by means of its lines of force, energy to the simple closed circuit coils on the isolated armature which enabled it to build up its own magnetic field that locked itself into the rotating magnetic whirlwind produced by the field coils. The need for a commutator was completely eliminated.

He worked out the design of dynamos, motors, transformers, and all other devices for a complete alternating-current system. He multiplied the effectiveness of the two-phase system by making it operate on three or more alternating currents simultaneously. This was his famous polyphase power system.

The mental constructs were built with meticulous care as concerned size, strength, design, and material; and they were tested mentally, he maintained, by having them run for weeks —after which time he would examine them thoroughly for signs of wear. Here was a most unusual mind being utilized in a most unusual way. If he at any time built a "mental machine," his memory ever afterward retained all of the details, even to the finest dimensions.

The state of supreme happiness which Tesla was enjoying was destined soon, however, to end. The telephone central station by which he was employed, and which was controlled by Puskas, that friend of the family, was sold. When Puskas returned to Paris, he recommended Tesla for a job in the Paris establishment with which he was associated, and Tesla gladly followed up his opportunity. Paris, he reasoned, would be a wonderful springboard from which to catapult his great invention on the world.

The budding superman Tesla came to Paris light in baggage but with his head filled to bursting with his wonderful discovery of the rotating magnetic field and scores of significant inventions based on it. If he had been a typical inventor, he would have gone among people wearing a look indicating that he knew something important, but maintaining absolute secrecy concerning the nature of his inventions. He would be fearful that someone would steal his secret. But Tesla's attitude was just the reverse of this. He had something to give to the world and he wanted the world to know about it, the whole fascinating story with all the revealing technical details. He had not then learned, and never did learn, the craft of being shrewd and cunning.

Six feet, two inches tall, slender, quiet of demeanor, meticulously neat in dress, full of self-confidence, he carried himself with an air that shouted, "I defy you to show me an electrical problem I can't solve"—an attitude that was consistent with his 25 years, but also matched by his ability.

Through Puskas's letter of recommendation he obtained a position with the Continental Edison Company, a French company organized to make dynamos and motors and install lighting systems under the Edison patents.

He obtained quarters on the Boulevard St. Michel, but in the evenings visited and dined at the best cafés as long as his salary lasted. He made contact with many Americans engaged in electrical enterprises. Wherever he could get a patient ear, among those who had an understanding of electrical matters, he described his alternating-current system of dynamos and motors.

Did someone steal his invention? Not the slightest danger. He could not even give it away. No one was even slightly interested. The closest approach to a nibble was when Dr. Cunningham, an American, a foreman in the plant where Tesla was employed, suggested formation of a stock company.

With his great alternating-current-system invention pounding at his brain and demanding some way in which it could be developed, it was a hardship for him to be forced to work all day on direct-current machines. Nowadays, though, his health was robust. He would arise shortly after five o'clock in the morning, walk to the Seine, swim for half an hour, and then walk to Ivry, near the gates of Paris, where he was employed, a trip that required an hour of lively stepping. It was then half-past seven. The next hour he spent in eating a very substantial breakfast which never seemed sufficient to keep his appetite from developing into a disturbing factor long before noon.

The work to which he was assigned at the Continental Edison Company factory was of a variegated character, largely that of a junior engineer. In a short time he was given a traveling assignment as a "trouble shooter" which required him to visit electrical installations in various parts of France and Germany. Tesla did not relish "trouble shooting" but he did a conscientious job and studied intensely the difficulties he encountered at each powerhouse. He was soon able to present a definite plan for improving the dynamos manufactured by his company. He presented his suggestions and received permission to apply them to some machines. When tested they were a complete success. He was then asked to design automatic regulators, for which there was a great need. These too gave an excellent performance.

The company had been placed in an embarrassing position and was threatened with heavy loss through an accident at the railroad station in Strasbourg in Alsace, then in Germany, where a powerhouse and electric lights had been installed. At the opening ceremony, at which Emperor William I was present, a short circuit in the wiring caused an explosion that blew out one of the walls. The German government refused to accept the installation. Tesla was sent, early in 1883, to put the plant in working order and straighten out the situation. The technical problem presented no difficulties but he found it necessary to use a great deal of tact and good judgment in handling the mass of red tape extruded by the German government as precaution against further mishaps.

Once he got the job well under way he gave some time to constructing an actual two-phase alternating-current motor embodying his rotary-magnetic-field discovery. He had constructed so many in his mind since that never-to-be-forgotten day in Budapest when he made his great invention. He had brought materials with him from Paris for this purpose and found a machine shop near the Strasbourg station where he could do some of the work. He did not have as much time available as he had expected, and, while he was a clever amateur machinist, nevertheless the work took time. He was very fussy, making every piece of metal exact in dimensions to better than the thousandth of an inch and then carefully polishing it.

Eventually there was a miscellaneous collection of parts in that Strasbourg machine shop. They had been constructed without the aid of working drawings. Tesla could project before his eyes a picture, complete in every detail, of every part of the machine. These pictures were more vivid than any blueprint and he remembered exact dimensions which he had calculated mentally for each item. He did not have to test parts through partial assembly. He knew they would fit.

From these parts Tesla quickly assembled a dynamo, to generate the two-phase alternating current which he needed to operate his alternating-current motor, and finally his new induction motor. There was no difference between the motor he built and the one which he visualized. So real was the visualized one that it had all the appearance of solidity.

The assembly completed, he started up his power generator. The time for the great final test of the validity of his theory had arrived. He would close a switch and if the motor turned his theory would be proven correct. If nothing happened, if the armature of his motor just stood still, but vibrated, his theory was not correct and he had been feeding his mind on hallucinations, based on fantasy not on fact.

He closed the switch. Instantly the armature turned, built up to full speed in a flash and then continued to operate in almost complete silence. He closed the reversing switch and the armature instantly stopped and as quickly started turning in the opposite direction. This was complete vindication of his theory.

In this experiment he had tested only his two-phase system; but he needed no laboratory demonstration to convince him that his three-phase systems for generating electricity and for using this current for transmission and power production would work even better, and that his single-phase system would work almost as well. With this working model he would now be able to convey to the minds of others the visions he had been treasuring for so long. He had just passed his twenty-seventh birthday.

Tesla now had available a completely novel type of electrical system utilizing alternating current, which was much more flexible and vastly more efficient than the direct-current system. But now that he had it, what could he do with it? The executives of the Continental Edison Company by whom he was employed had continually refused to listen to his alternating-current theories. He felt it would be useless to try to interest them in even the working model. He had made many friends during his stay in Strasbourg, among them the Mayor of the city, M. Bauzin, who shared his enthusiasm about the commercial possibilities of the new system and hoped it would result in the establishment of a new industry that would bring fame and prosperity to his city.

The Mayor brought together a number of wealthy Strasbourgers. To them the new motor was shown in operation, and the new system and its possibilities described, by both Tesla and the Mayor. The demonstration was a success from the technical viewpoint but otherwise a total loss. Not one member of the group showed the slightest interest. Tesla was dejected. It was beyond his comprehension that the greatest invention in electrical science, with unlimited commercial possibilities, should be rejected so completely.

M. Bauzin assured him that he would undoubtedly receive a more satisfactory reception for his invention in Paris. Delays of officialdom in finally accepting the completed installation at the Strasbourg station, however, postponed his return to Paris until the spring of 1884. Meanwhile, Tesla looked forward with pleasurable expectancy to a triumphant return to Paris. He had been promised a substantial compensation if he was successful in handling the Strasbourg assignment; also, that he would be similarly compensated for the improvements in design of motors and dynamos, and for the automatic regulators for dynamos.

When he got back to the company's offices in Paris and asked for a settlement of his Strasbourg and automatic-regulator accounts, he was given what in modern terminology is called the "runaround." To use fictitious names, as Tesla told the story, the executive, Mr. Smith, who gave him the assignments, now told him he had no jurisdiction over financial arrangements; that was all in the hands of the executive, Mr. Brown. Mr. Brown explained that he administered financial matters but had no authority to initiate projects or to make payments other than those directed by the chief executive, Mr. Jones. Mr. Jones explained that such matters were in the hands of his department executives, and that he never interfered with their decisions, so Tesla must see the executive in charge of technical matters, Mr. Smith. Tesla traveled this vicious circle several times with the same result and finally gave up in disgust. He decided not to renew his offer of the alternating-current system nor to show his motor in operation, and resigned his position immediately.

Tesla was undoubtedly entitled to an amount in excess of

$25,000 for the regulators he designed and for his services in Strasbourg. Had the executives been endowed with even a smattering of horse sense, or the ordinary garden variety of honesty, they would have made an attempt to settle for $5,000, at the least. Tesla, hard pressed for cash, would undoubtedly have accepted such an amount, although with a feeling that he was being cheated in a large way.

Such an offer would probably have held Tesla on the payroll of the company and preserved for it the possession of the world's greatest inventor and one who at the time had definitely demonstrated he was an extremely valuable employee.

One of the administrators of the company, Mr. Charles Batchellor, Manager of the Works, who was a former assistant and close personal friend of Thomas A. Edison, urged Tesla to go to the United States and work with Edison. There he would have a chance to work on improvements to the Edison dynamos and motors. Tesla decided to follow Mr. Batchellor's suggestion. He sold his books and all other personal possessions except a few articles which he expected to take with him. He assembled his very limited financial resources, purchased tickets for his railroad trip and transatlantic journey to New York. His baggage consisted of a small bundle of clothes carried under his arm and some other items stuffed into his pockets.

The final hours were busy ones and, as he was about to board the train, just as it was ready to pull out of the station, he discovered his package of baggage was missing. Reaching quickly for his wallet, which contained his railroad and steamship tickets and all his money, he was horrified to discover that that too was missing. There was some loose change in his pocket, how much he did not know—he did not have time to count it. His train was pulling out. What should he do? If he missed this train, he would also miss the boat— but he could not ride on either without tickets. He ran alongside the moving train, trying to make up his mind. His long legs enabled him to keep up with it without difficulty at first, but now it was gaining speed. He finally decided to jump aboard. The loose change he discovered was sufficient to take care of the railroad fare, with a negligible remainder. He explained his situation to the skeptical steamship officials and, when no one else showed up to claim his reservations on the ship up to the time of sailing, he was permitted to embark.

The ship offered little to interest him. He explored it thoroughly and in doing so made some contacts with members of the ship's company. There was unrest among the crew. There was unrest in Tesla also. He extended sympathy to members of the crew in their claimed unjust treatment. The grievances affecting the crew had built up one of those situations in which a small spark can cause a large explosion. The spark flew somewhere on the ship while Tesla was below decks in the crew's quarters. The captain and officers got tough and, with some loyal members of the crew, decided to settle the trouble with belaying pins as clubs. It quickly became a battle royal. Tesla found himself in the middle of a fight in which when anyone saw a head he hit it.

Had Tesla not been young as well as tall and strong, his useful career might have ended at this point. He had long arms in proportion to his six feet, two inches of height. The fist at the end of his arm could reach as far as a club in the hands of an adversary, and his height enabled him to tower over the other fighters so his head was not easy to reach. He struck hard and often, never knowing for or against which side he was fighting. He was on his feet when the fight was over, something which could not be said of a score of the crew members. The officers had subdued what they called a mutiny, but they too carried indications that they had been through a battle. Tesla was definitely not invited to sit at the captain's table during the voyage.

He spent the remainder of his journey nursing scores of bruises and sitting in meditation at the stern of the ship, which too slowly made its way to New York. Soon he would set foot on the "land of golden promise."

The engineer has worked side by side with the general from the earliest days of warfare. Sometimes, as in the case of the catapult, the battering ram, the machine gun, and the buzz bomb, his contribution has been purely military in nature. More often, it has been, in essence, general engineering applied to military purposes. Bridges and roads must be built and frequently destroyed. Complex electrical systems for the firing of guns and rockets must be devised. Atomic submarines which can submerge indefinitely must be constructed.

The contributions of Archimedes and Leonardo to the art of destruction have already been noted. The growth of scientific engineering knowledge has of course revolutionized warfare. From the bridge over the Hellespont which Darius used in his invasion of Greece the story is one of ingenuity, often under the most hazardous circumstances. A single example, the construction of the tunnels at Messines which decided a crucial battle of World War I, is related here.

TRIUMPH AT MESSINES

LEON WOLFF

SIR HERBERT PLUMER'S HAIR had turned white during his two thankless years as warden of the salient. A heavy responsibility had been his, with no chance for glory, for there was hardly a point within the loop of ground held by his Second Army which German guns could not enfilade or fire into from behind—a state of affairs hardly calculated to improve the nerves of this commander or his troops. Nonetheless he had made of the salient a nut so hard to crack that the enemy had not tried to do so since 1915. An ideal officer to hold any position in bulldog fashion, Plumer was a prim little old man with a pink face, fierce white mustache, blue eyes, a little pot-belly mounted on tiny legs.

He was fortunate in possessing an extraordinary chief of

staff, the cultured and wise Major General Sir Charles ("Tim") Harington—also a cautious planner, but with, perhaps, an extra dash of imagination and verve. He was tall and thin, nervous, had a card-index mind and a sense of humor. The combination of the two men had proved outstanding in the war to date; and now they hoped to prove that they could storm an objective as well as hold one. Surely they had been allotted more than enough time to get ready. During the long, lean years while they had hung on, improving their defenses, not looking for trouble, they had studied their terrain with microscopic thoroughness. Of Plumer it was said that "he knew every puddle in the salient." Plans for capturing the Messines-Wytschaete ("White Sheet") flank of the ridge had been endlessly cast and recast in conference after conference, order after order. By spring 1917 the operation had been worked out with an intensity unmatched in the war thus far.

Plumer's trump card was a system of enormous land mines burrowed beneath the German front. This work had begun in 1915 with the construction of shallow galleries and small charges about fifteen feet underground. Next year the idea of concentrating on a deep mining offensive, with tunnels and charges nearly 100 feet below the surface, was contemplated. But to penetrate secretly with galleries of substantial size the saturated, semiliquid layer that made up the Flanders subsoil was on the face of it quite impossible. Might there be another way?

The problem was studied by Lieutenant Colonel T. Edgeworth David, chief geologist of the BEF, and the engineer in chief, Brigadier General G. H. Fowke, who analyzed sand and clay layers and the variations of water in each. Perhaps, they thought, the layer of heavy blue clay lying even farther down—between 80 and 120 feet—might be a practical medium for their purpose. At that depth the tunnels and charges could not be blown up, accidentally or otherwise, by mortar fire or shallow countermines; and the sound of digging would be so muffled that secrecy might be possible. Under the harried conditions of war, with the time element so important, and specialized heavy equipment so scarce, was it possible to construct such shafts, to lay gigantic charges accurately under the key German frontline positions, within a reasonable time, and without being detected? Though the odds were not good, it was decided to try.

By January, 1916, six tunnels had been started (the signs above them read "Deep Wells"); and during the next year twenty of the largest mines in the annals of warfare were in place or in process of being placed. Twenty underground communities came into being. From the sandbagged openings wooden stairs led down to sleeping quarters. Below these, planked passages slanted to headquarters posts and thence to the actual three-by-six galleries. In these bowels of the earth, the molelike character of the war was fantastically intensified. The hum of pumping engines never ceased. The thousands of men who worked here with picks and shovels, coughing in the dampness, white of skin, shoved their tunnels a pitiful ten or fifteen feet forward each day under the glare of electric lamps; and as each gallery was completed a mine was laid in place (sometimes two) containing charges of ammonal up to 95,000 pounds. Some tunnels were almost half a mile in length. By June 7th the total had come to a million pounds of explosive and almost five miles of gallery.

Meanwhile the Germans were also mining toward the British lines, but in a smaller way and much more shallowly, for they had neither the equipment nor the plan for deep works possessed by their enemy. Yet in places they did venture to considerable depths—sometimes nearly 60 feet down—and many was the time when the British, listening with microphones at the forward, boarded-up faces of their galleries, heard with dismay a German tunnel approaching their own. Early in 1917 the enemy had dug within eighteen inches of the British at the northern corner of the ridge.

A colonel came to Harington with the news and recommended that the mine be blown. Harington thought for a moment, walked to Plumer's door, knocked, and entered.

" 'Mines' says we must blow the Hill 60 mines today."

"I won't have them blown," snapped the general. "Good night."

Work stopped, the British evacuated the tunnel, and by chance the Germans veered away.

The enemy knew that the ridge would soon be attacked. Preparations above ground were obvious, and they had captured many prisoners who had talked. One specifically told them on May 29th that the assault would begin June 7th after eight days of bombardment. But what about the mines, the only element of real surprise? The Boches had their suspicions. While they did not greatly fear the small, shallow charges they had become apprehensive concerning deep mining. They sent out many raiding parties primarily to bring back, not prisoners, but samples of the soil thrown up by diggings. On April 9th one of these parties returned with blue clay. While this was a sure sign that the British were constructing at least one mine at great depth, the Germans reacted inconsequentially, assuming evidently that the shaft was only an isolated one (if it existed at all), and that, since the attack was so imminent, little further mining could be accomplished by the British in the brief time remaining.

To the British it seemed doubtful that their opponents could not know of the twenty deep mines. Certainly, they thought, some prisoners had disclosed them by now. (Not one had done so.) Surely the Germans had been able to hear the work going on, despite many soundproofing measures. (They had not, due to the inferiority of their microphones.) For once, even Sir Douglas Haig seems to have been troubled by nervousness. His greatest fear was that the enemy might abandon his front lines just before the attack, a suggestion in fact made by Lieutenant General von Kuhl, Crown Prince Rupprecht's Chief of Staff, early in May. But this projected withdrawal, which would have dislocated Haig's plan even worse than the one which upset Nivelle, was rejected by Rupprecht. The blow to morale would be too great, he estimated, if such an outstanding defensive sector were to be discarded without a fight. It could and would be held. As for British mining, the Germans had rather decided by May that it had ended, except for small efforts of no importance. Only at Hill 60 was it definitely known that the British were still digging in earnest, and here, according to the German officer in charge, their work had been hopelessly damaged by countermining. Thus misled by faulty Intelligence, the Crown Prince could not see how the clear and aboveboard preparations of the enemy could possibly succeed.

Accordingly, on June 1st he caused the XIX Army Corps to issue an order concerning Wytschaete and Messines:

> These strong-points must not fall even temporarily into the enemy's hands. . . . They must be held to the last man even if the enemy has cut them off on both sides, and threatens them from the rear.

Furthermore, the German troops were told that they need have no fear of a breakthrough. Reserves were already in place and would move in swiftly to seal off any gaps that might occur. In these orders and reassurances the opposing high command clearly showed the value they placed on the east and south ridges of their precious encirclement.

So Haig's one great worry was baseless; the Germans would stay put and make their stand. Not knowing this, he recommended in a conference on May 30th that all mines be exploded before zero day; next the troops would occupy the ground; and later they would try to cross over the crest of the ridge. He also suggested moving the target date forward a day or two. Plumer begged to reject these last-minute changes, and Haig agreed to let matters go forward on schedule.

The trepidation of the British is understandable. The greatest series of simultaneous explosions in history was about to take place (it would triple the former record set in New York during subway construction late in the 1800's). Hundreds of thousands of men had been working toward this one day for over two years. The immensity, the importance of the operation was incalculable. And it all hung on the feeblest of threads. One British private soldier taken prisoner could have nullified it. German detection devices could have been alert to the entire plan for months. Airplane observation might have detected any number of blue-clay diggings, despite efforts to camouflage them as they were hauled to the rear. How much did the Germans know? From Haig on down, the British would have sold their souls for an answer.

The very fact that German countermining continued was noteworthy, even though most of these efforts were at shallow levels. In one place the Germans were known to be digging along a line that was bound to intersect the British gallery. Again this was near (or rather under) Hill 60, where the most spectacular mining, countermining, and mine fighting took place. The state of affairs was later described:

> ... on May 9th the enemy was so near that work was stopped, and the branch gallery was loaded with 1600 pounds of ammonal. The Germans had evidently completed their shaft and were driving a gallery past the end of the branch gallery. As, however, there was only a month to go, and the *camouflet* (a small defensive charge which ruins the enemy tunnel but does not open the surface of the ground) might detonate the great mine, or at least cause the Germans to probe vigorously, it was decided that the safest course was to accept the risk involved in letting the enemy work on, and not to fire the mine unless he touched the actual timbers of the branch gallery. ... The Germans could now be heard putting in timber, working a truck, walking, and even talking. On May 25th in some other workings they fired a mine whose position was "dangerously correct" directly above the Hill 60 gallery. It crushed in the junction of the galleries and entombed two listeners. One, Sapper Earl, in the Hill 60 gallery, coolly went on listening and heard a German walk down an enemy gallery apparently directly over the great mine. ... The listeners had to be withdrawn, and from then onwards the staff could only trust that the enemy would not reach the British workings before the mine was fired.

At Petit Douve, near Messines, one mine, already laid and electrically charged, was discovered by the Germans. They blew a *camouflet*, wrecked the main British gallery, and the mine had to be abandoned. This left nineteen.

By late evening, June 6th, these nineteen mines were in

place and charged, their shafts were tamped down, and the only remaining question was how many of them—especially the old ones that had been laid down as long as six months ago—would actually explode next morning.

The mines were to be only a curtain-raiser. Because they were an unknown factor to large degree—nobody really knew how many would go off, whether each lay in exactly the right position, and how much damage they would do to men, trenches, and guns many yards above—the greatest artillery mass of the war had been arrayed against enemy lines between Ploegsteert ("Plug Street") Wood and Observatory ridge, about a mile northeast of Hill 60. Over 2400 guns and howitzers were to participate, fully a third of which were heavy pieces: one gun to every seven yards of front.

Other than the mines, as we have seen, there would be no surprise at all (except possibly for the precise moment of zero hour itself, and this was not officially told to the troops until the 5th); and the artillery preparations were unusually brazen. In single file the heavies were hauled directly to the frontal area from the rear towns and assembly points. Behind them jostled the little field guns, galloping up without the slightest caution, a wild noisy collection followed closely by their ammunition wagons. They were emplaced wheel to wheel, with no attempt to hide them.

In the final days gas shells were thrown in vast numbers to force the enemy to don masks and lose sleep. And, further to confuse him, the bombardment was twice increased to pre-attack intensity, and twice the Germans reacted spasmodically to false alarms.

Tanks, too, were in readiness—seventy-two Mark IV's that assembled a few miles southwest of Ypres and waited under camouflaged shelters for the signal to proceed toward the front. The night of the 6th they emerged, throbbing and clattering, and approached their starting points under the cover of airplanes which flew back and forth to drown out their noise.

Three hundred planes of the II Brigade Royal Flying

Corps went into action late in May, mostly to assist the artillery by observation and photographs.

The attack itself was to be a straightforward operation along a ten-mile front toward a final objective two miles away at most, known as the Green Line, or Oosttaverne Line, running slightly to the east of that former village in a nearly straight line that formed a chord across the base of the German salient. Three corps (IX, X, and II Anzac) would participate with three divisions each, and each corps would have one division in reserve, ready to leapfrog through upon signal. About 80,000 infantry would go over the top at dawn, at which moment the mines would be detonated and the artillery barrage would commence with every operable gun along the Second Army Front.

The men waited, worked, and trained. For once, in World War I, they approached zero hour with a sense of optimism, though they understood what they faced. "The enemy will fight his hardest for the Messines ridge," said an officer. "He has stacks of guns against us."

And another, who knew the salient perhaps too well, peered at the German-held ridges, lofty and arrogant in the hazy distance, flaming with guns, webbed with row after row of sandbagged trenches, peppered with thousands of machine-gun emplacements, pillboxes, and sharpshooters' nests. He turned to a newspaper man and murmured, "It's a Gibraltar." The mines would have to do their job.

Major General Harington opened his advance press conference with these words: "Gentlemen, I don't know whether we are going to make history tomorrow, but at any rate we shall change geography."

During the evening the men marched silently in columns of fours like groping tentacles toward the communication trenches, and thence to the front, where white jumping-off tapes lay on the soft wet ground of No Man's Land. They were troubled and wearied by the need for wearing their masks, for gas shells were plopping all about them, laying low the unwary and careless as well as many pack animals gasping and heaving in the poisoned air. It was warm that night. Fog lay on the salient like a heavy caress, and in it not a breeze stirred. Overhead forked lightning played, accompanied by the mutter of thunder. At midnight a sharp thundershower broke. It lasted only a few minutes, and after it passed, a three-quarter moon floated regally in a nearly clear sky. Now brilliant flashes against the enemy slopes could be seen, and the steady whamming of the big guns sounded perceptibly louder as the blanket of fog melted away.

A half-hour before zero hour the British guns stopped firing, and the night became so still that one could hear nightingales singing in the nearby woods. The men fixed bayonets and removed their gas masks. Some of them dozed. Officers changed to enlisted men's tunics and kept peering at their wrist watches. Zero hour would be 3:10.

At 2:52 the Germans threw up yellow and green flares, calling for artillery fire—a disconcerting sign. How much did they know? At 2:57 heavy bursts of shrapnel swept segments of the British front; but quickly it ceased.

At 3:05 the first streaks of dawn filtered over the Messines ridge. On Mount Kemmel the cocks began to crow. Two green star-flares burst directly in front of the New Zealand division at 3:06, then machine guns and another flare. Had this unit been discovered? (Some of their assembly trenches had been dug dangerously forward in No Man's Land during the early evening.) But at 3:09 the German guns stopped chattering. For one minute absolute silence saturated the air while at whispered orders the troops crawled over their parapets and lay flat in front of the tapes.

A few seconds before 3:10 some of the heavy guns rearward began to fire. Then each of the nineteen land mines exploded almost in unison. The earth quaked, tumbling and staggering the British soldiers as they rose in awe to see the rim of the hated ridge burst skyward in a dense black cloud, beneath which gushed nineteen pillars of flame that lit the salient with the red glare of hell. The pillars fused into greater mushrooms of fire that seemed to set flame to little clouds above. Then, a moment or two later, the long roar of nineteen explosions blended and reverberated into one long blast that stunned even the British troops, awakened the countryside, rolled through Flanders and northern France, hurtled the Channel, and was heard in London by Lloyd George, awake in his study at Number 10 Downing Street.

From the German positions yellow flares soared imploringly high into the sky, the pathetic prayers of doomed men crying for help. As the villages of Messines and Wytschaete disappeared into oblivion, the heaviest of all artillery barrages struck the German front, and the British assault brigades scrambled over the top.

Plumer and his staff had breakfasted at 2:30, after which everybody went to the top of Cassel Hill near Second Army headquarters, a few miles deep in the salient, to see the mines go up—all but Plumer, who returned to his room and knelt at his bed in prayer. When the first news of the infantry advance came, he burst into tears. It was quite clear that his Second Army was winning as planned, and with greater ease than had been expected. So swift and thorough was the British success that subsequent fighting came as an anticlimax.

The enemy was in a state of near shock when the British fell upon them. They surrendered en masse, weeping, waving handkerchiefs, grasping the ankles of their captors. Thousands lay beneath the ground, to be forever entombed there. Some of the mine craters were three hundred feet across and seventy feet deep. The wreckage of their front left many Germans cringing in derelict shelters "like beaten animals" while the British walked along throwing Mills bombs at unresisting clusters of men too dazed to surrender. One Australian lieutenant reported how they "made many fruitless attempts to embrace us. I have never seen men so demoralized." Another distraught prisoner said that only two men in his section of the line had survived the blast. A captured officer reported that of his two-hundred-man company only thirty were alive when the British foot soldiers arrived. The 3rd Bavarian Division was relieving the 24th Saxons precisely when the explosions and the attack burst; both relieved and relievers were decimated, and most of the balance were made prisoners.

Certainly the operation had been, in the words of one writer a "siege-war masterpiece," one "in which the methods employed by the command completely fitted the facts of the situation," a triumph of engineers in what was essentially an engineers' war.

IV. Toward The Future

IV. Toward the Future

Barring military catastrophe, the society of the future will be increasingly conditioned by the engineering profession. In Samuel Lilley's article on automation, we have seen how the unskilled worker is gradually being eliminated and how even the moderately skilled will eventually be replaced by the engineer. In Sir Christopher Hinton's article on atomic power, we see the implications of almost unlimited power for the future conduct of human affairs. The dreams of the engineer have become more all-encompassing than his most useful past achievements. The present section offers three typical examples.

"Abu Simbel" by Ritchie Calder spans the whole history of the profession, from Egypt's master builders to the problems created by the Aswan High Dam now being constructed. It shows how the most modern engineering methods will preserve one of the great engineering monuments of ancient times. The servants of the Pharaohs carved a temple out of a mountain, using skills few moderns realize they possessed. Now the entire mountain with the temple it encloses will be lifted intact a distance of two hundred feet. The present article, written for the "Unesco Courier," has helped focus world attention on the project. The author has written numerous articles on scientific subjects for the British press and is the author of a number of excellent books for the layman, including The Inheritors: The Story of Man and the World He Made and The Profile of Science.

ABU SIMBEL

RITCHIE CALDER

UNESCO HAS ADDRESSED an appeal to its one hundred member nations for voluntary contributions to save the two colossal temples of Abu Simbel from destruction under 200 feet of water as a result of the construction of the High Dam at Aswan. The temples, hewn out of the solid mountain rock in the thirteenth century B.C., are considered to be among the great architectural glories of Pharaonic art.

The two temples of Abu Simbel are to be preserved by what has been described as "the most daring engineering project of modern times." They will be sliced from the mountainside into which they were built three thousand years ago, encased in concrete boxes, and raised by giant jacks over two hundred feet.

Four miles south of the present Aswan Dam in central Egypt, the new High Dam (Sadd el Aali) is being built across the Nile by the United Arab Republic with Soviet aid. It will contain the river above that point and form a gigantic artificial lake drowning the Nile valley of Nubia in both Egypt and the Sudan.

For Egypt the new lake will mean the beginning of an era of great regeneration. It will increase the total food production of the country by nearly one half; some 2,500,000 acres of desert will be brought under cultivation, and an additional 750,000 acres now flooded will be reclaimed. Egypt's hydroelectric output will be increased by the dam something like ten times.

The new dam is an economic necessity for Egypt. But there remains the fact that everything within the Nubian lake area will be obliterated by 1968. In 1959, the Governments of the U.A.R. and the Sudan turned to UNESCO for help to save the temples and monuments menaced by the impending flood in both countries. On March 8, 1960,

UNESCO's Director General, Mr. Vittorino Veronese, launched the now famous International Campaign to save the treasures of Nubia.

The enormous tasks of the campaign were divided into three categories. First there was the urgent job of carrying out systematic surveys and the organization of expeditions to excavate archæological and prehistoric sites in the threatened area of both Egypt and the Sudan.

The second category comprises the dismantling and the removal of ancient temples and other historic monuments to safety beyond the flood area. These operations are now under way and are progressing satisfactorily.

The last category includes the two most important projects of the campaign—the preservation *in situ* of the temples on the isle of Philae, and the Abu Simbel temples.

The greatness of Egypt is popularly identified with the petrified geometry of the pyramids, those remarkable examples of civil engineering, veritable mountains of masonry. Experts, however, would agree that the sublimation of funerary art, combined with shrewd scientific insight, were the rock temples.

There are great free-standing Egyptian temples, superb in their proportions, in their pillars and their sculptures, but the architects of these were able to choose the best artificial site, and the masons to select or discard the quarried blocks for their building and ornamentation. A sculptor, carving a statue, would pick the most suitable, most enduring and most flawless stone for his purpose.

The architects and the masons of the rock temples had no such latitude; instead of choosing blocks of stone they had to discover an escarpment or a mountain which would conform to their exacting requirements. Once committed, their artistic ingenuity was hostage to the site.

Among the greatest of these temples hewn from the living rock were the Great Temple and the Small Temple of Abu Simbel in which the immortality of the Gods and of Rameses II was embodied in the indestructible rock. They were planned by Seti I but executed in all their grandeur by Rameses II during his prodigious reign of 67 years, from 1300 to 1233 B.C.

Modern geologists who have examined the site of Abu Simbel are unstinted in their tribute to their unknown predecessor (or predecessors) who chose this particular location 3200 years ago.

Thirty miles north of the Second Cataract, on the left bank of the Nile, where the river turns east, were two rocky prominences divided by a gully. Here was a site which met a first elementary requirement: that the temple, dominating the river, face the rising sun.

In the vicinity, the left bank of the valley is steep, a cutting by the river exposing about 400 feet of sandstone cliff, which *could* have shown the ancient geologists the structure of the rocks. But this evidence was merely superficial.

How did they know, as can be determined nowadays, that there was little or no distortion of those sandstone beds? How did they establish that the interior of the mountain could architecturally house the Great Temple which they were proposing to excavate? How did they satisfy themselves that the consistency of the sandstone would lend itself to the carving of colossi and of the friezes?

How much did they know about the chemistry of minerals and how the elementary granules of the sand were bound together by a cement of iron oxide, which gives the rock its color gradations, ranging through all shades from pink to dark mauve? What did they know of the porosity of the rock and the highly solvent power of the Nile water, and hence of the water-table below the mountain, which, when the rocks were exposed to heat, would be pumped upwards by capillary action? This "pumping" would mean the dissolving of minerals in the rocks, a chemical reaction, and the precipitation of salts—all likely to alter the characteristics of the rocks.

The carved immortality of their Pharaoh would depend upon the durability of the rocks they found. How much did they know about weathering? How did they determine from external evidence that the rocks within the mountain would lend themselves to the structural engineering of an ambitious temple such as Rameses would certainly demand? On decisions like these not only the reputations of those ancient geologists and engineers depended but, one would guess, their very lives.

To quote one of the technical reports drawn up by engineering experts:

> The temples of Abu Simbel are a wonderful achievement. Apart from the importance of the monuments themselves, we are struck with admiration at the deep knowledge of geology which the ancient Egyptians possessed. The presence of hard sandstone banks alternating with softer ones was used to advantage in creating the temples and the statues. The more compact layers were chosen for the ceilings of the temples and inner rooms, or to support the greater weight of the sitting statues. They also made the most of the fissures in the rock: the façades of the two temples run parallel to the more fissured lines.

On what we would nowadays call the "feasibility report," the ancients went ahead with the construction of the two temples—some 300 feet apart—overhanging the banks of the Nile. But they did something more; they contrived the design of the temples to fit into the natural landscape so that art and science conspired with nature to make Abu Simbel one of the wonders of the world.

The larger of the two rock temples, facing to the east and to the rising sun, has a façade 33 meters (over 107 feet) in height and 38 meters (over 123 feet) in width. It was dedicated by Rameses to Ra-Horakhti, Amon-Ra, and Ptah, the most important gods of Egypt. In its proportions and structure it equalled a temple that might have been built on the surface instead of underground.

To Nefertari, his wife, Rameses dedicated the Small Temple, several hundred feet away across a sandy gully. Its façade measures 88 feet in width and 39 feet in height. It is ornamented by six colossal statues, each 33 feet high. They are in two separate groups in each of which the queen stands between figures of the Pharaoh with, at their feet, their sons. One of the pillars states: "The King built this Temple by hewing it from the rock of the hill of the country of Takens." For the preservation of the Abu Simbel monuments, four proposals were given serious consideration:

1] Raising the temples and surrounding rocks above the level of the waters;
2] Building an earth and rock-fill dam, to protect both temples;
3] Building a protective concrete dam in front of each temple,
4] Building a large dome concrete dam, to protect both temples.

Of all these possibilities for preserving the temples from the flood, two schemes remained at the end of 1960. One was a proposal submitted by a French engineering concern, the Bureau d'Etudes André Coyne et Jean Bellier (Paris), to build a rock-fill dam to enclose the two temples. The size was planned to allow a clear space in front of the temples but the dam itself would tower above the façades and, among other things, would defeat the ingenuity of those ancient engineers who contrived that the rays of the rising sun would penetrate into the heart of the mountain.

Technical difficulties of water seepage would have meant, in perpetuity, an annual cost of $370,000 for pumping. The capital cost of the dam construction would be $82,000,000. Another objection raised was that, with the filling of the High Dam, the water-table of the land around would rise and that, even though enclosed by the protecting dam, the rocks of the Abu Simbel mountain would become saturated and the seepage would affect the temples imbedded in it.

The second proposal was of a very different kind. Conceived by the Italian architect and archæologist, Piero Gazzola, it was prepared by three Italian civil engineering firms —Italconsult, Impresit, and Lodigiani and put forward to UNESCO by the Italian Government in October 1960. This suggested encasing each of the two temples in concrete boxes and raising the rock masses 60 meters (200 feet) above the present level of the waters and reconstructing the original hill landscapes so that the final position of the temples would bear the same relationship to the Nile as it does at present—but on a higher elevation.

In January 1961, a Committee of Experts appointed by UNESCO in conjunction with the Government of the United Arab Republic unanimously recommended the acceptance of the Italian project for lifting the temples.

The Italian scheme is bold, imaginative, and grandiose and has transferred into twentieth-century idiom something of the ancients' massive thinking.

The two temples will be lifted separately and each will be supported by a honeycomb of concrete supports, forming two massive pedestals. The lifting will be an engineering feat, raising, in the case of the Great Temple, a wedge of mountain rock, weighing a quarter of a million tons, with a delicacy of movement never before attempted.

Consider again the site. The two temples are imbedded in solid rock. The façades with their massive statuary are poised on the escarpment above the river. The platform-rock on which they stand is 120 meters (390 feet) above sea level. When the High Dam is completed, the water will rise to 182 meters (591 feet) above sea-level. So to be above water-level the rock sections embodying the temples will have to be raised at least 200 feet, and underpinned.

Before the lifting operations begin, a whole series of precautionary measures will have to be taken. The area in front of the temple will have to be protected by a coffer-dam, an embankment 134 meters (440 feet) high, so that the operations can be protected as the dam waters rise. A pit as long as the temple façades will be dug down to the 105 meters level, that is, about 50 feet deep, from which the excavation and clearing of the galleries beneath the temple blocks will proceed.

At the same time, the temples themselves will be carefully reinforced—an operation almost as delicate as the mending of cracked porcelain. The structures, including the colossi, have already been carefully examined for fissures.

Some of the cracks probably existed when the temple was first built. They have caused damage to the Ramessid colossi in front of the Great Temple—and in the case of one statue have led to the shearing off of a head and arm.

Inside the temple there are natural pipes, hollowed out by percolating water, and veins which were closed with mortar when the temple was built. There is the risk of scaling-off of weathered rock when work begins. Of the eight columns in the main hall of the Great Temple, two are not carrying any load because of cracks between the roof and the column head, and two others have been weakened by fractures.

The walls and roofs are in pretty sound condition but the greatest precautions will have to be taken to preserve intact the inscriptions and painting. (This could be done by covering them with adhesive textiles.) All this work, which would have amounted to a major piece of restoration even if the temples were not due to be moved, will have to anticipate any disturbance of the rock blocks in which the temples are encased.

The proposal is that the blocks, and the temples they contain, once cut out of the hill, should be enclosed in a reinforced concrete box so rigid that the separation of one block from the surrounding rock and the subsequent lifting will not introduce any stresses or strains. This box would have a bottom and four sides and binding ties across the top.

The bottom is a matter of special engineering concern because it will have to take the lifting forces. The proposed bottom grid structure will be $16\frac{1}{2}$ feet high (this may be reduced to 13 feet), which will involve the excavation of three groups of five parallel tunnels. The side and rear walls of the box will be of heavily reinforced concrete. They will be constructed in three groups of vertical shafts, excavated and concreted, and each wall section will be rigidly connected to the bottom girders.

The front walls have to take care of the façades of each of the temples. The space between the wall and the façade will be filled with compacted, crushed sandstone—like packing a fragile present for dispatch.

But, first, the mountain above the rock blocks has to be removed, to reduce the weight and facilitate operations. This will mean shifting the overlay from 155 meters (about 500 ft.) above sea level upwards. This may sound like nothing more than a rather elaborate piece of landscape gardening (since it is planned to restore the contours around the lifted temples), but it actually is an extremely tricky operation. Explosives cannot be used nor any method which might cause vibrations likely to disturb the temple rocks.

That applies to the whole process of cutting out the blocks. Apart from explosive shocks being entirely barred, any tools which would cause serious vibrations would be dangerous. However, experiments carried out in Norway, on rocks which would vibrate more than the Nubian sandstone, have shown that compressed air tools for boring and breaking can be safely used.

In rocks within 24 feet of any decorated parts, mechanical hammers weighing not more than 30 kilograms and giving a thousand blows per minute may be used. At closer range electric saws and chisels will have to be used.

But before anything is done, experiments will be carried out reproducing similar conditions to those at Abu Simbel. And the experts have recommended that during the operations microseismic warning systems—supersensitive earthquake meters—should be installed. In effect, it will be a matter of the doctor feeling the patient's pulse the whole time, one of the "patients" in this case weighing a quarter of a million tons!

Once the blocks are cut and boxed the most delicate operation of all will begin. This is the edging upwards, two millimeters at a time, of the concrete "crates" and their massive contents. This will be done by mechanical jacks—on the same principle as those used to hoist a car when you have a punctured tire. Outsized versions of those will be installed within the grid of the bottom structure in a lattice so contrived that the points of upward pressure will be evenly distributed.

The original proposal was to use hydraulic jacks. A later suggestion was for mechanical jacks with a maximum lifting capacity of 2,000 tons each, driven by hydraulic stroke cylinders moving a turnable nut through a ratchet system. The number of jacks and their lifting power are still a matter of debate among the experts. Some want more of them —250 with 1,000-ton lift each and even duplication of that, so that one set of jacks would always be taking the load while the other was being retracted.

Although the edging upwards will be done less than $\frac{1}{16}$ of an inch at a time, one complete lifting cycle will raise the temple by one foot. At this stage of the lift, prefabricated reinforced concrete blocks of 30 centimeters will be slipped into position to form permanent supporting pillars. All this will be carefully synchronized and the functioning of each jack will be recorded on a central control panel so that the chief supervisor can make sure that there is no risk of warping, which would produce stresses in the temple blocks.

As the blocks rise upwards, the jacks will follow on taller and taller concrete pillars which will form part of the final supporting structure. The stability of all this stilt-like erection will be guaranteed by horizontal reinforced concrete beams joining the continuous walls. Thus in the end, each temple will be standing on the crest of what will, externally, be an enclosed pillar, surely the biggest of pedestals which museum exhibits have ever had.

This, the direct engineering process, does not take into account the vast "housekeeping" program involved. Already surveys have discovered where the huge supplies of suitable sand and aggregate can be obtained in the neighborhood.

There will be a floating quay, which will rise with the waters of the Nile. There will be a network of supply roads; an electricity generating station; a settlement with a working population, and with civilized amenities which will be desperately needed in this tract of desert; workshops, etc., commensurate with one of the biggest engineering enterprises; and when the "surgical" operation of cutting out and transplanting the temples to the point where they will be safe from the encroaching waters of the High Dam of Sadd el Aali is performed, there will be the further job of "plastic" surgery.

The "scalped" mountain will be restored and the features, even the parting, the gully, between the temples will be reproduced, as a replica, on the higher level, of what they are today. This is not merely "beauty treatment," or even æsthetics, although those are properly commendable; it is also a question of the future preservation of the monuments.

We talk about "living rocks," transferring a biological simile to geology; but it is certainly true that these monuments have survived with remarkable durability because they have "lived" in balance with their natural surroundings.

The present day experts have a healthy respect for the foresight and judgment of their predecessors of 3,200 years ago. The Ancients chose the site well, for all the reasons which have already been discussed, but they also chose an environment in which exposed masonry could endure. This involved factors of temperature, humidity, shelter from the sand blast, of erosive desert wind, and a subtle communion with the waters of the Nile. (Only since the building of the original Aswan Dam with the resulting raising of the water has chemical disintegration, in the base of the Small Temple, begun to appear; this mineralogical gangrene will be halted by the proposed "surgical" operation.)

When the dam is filled the Nile will be approximately in the same relationship to the temples as it is at present, and at the higher elevation, contours similar to the present ones will recapture the environment. The orientation of the temples will be precisely as now and the morning sun will continue to bring out of the darkness the faces of the Ancient Gods.

Willy Ley was a founder of the German Rocket Society and one of the pioneering writers on rocketry and space travel.
He now resides in the United States where his books on scientific subjects, such as Rockets, Missiles and Space Travel, have won wide popularity. In 1954 he published Engineer's Dreams, a description of nine engineering projects for the future. These projects partake of many of the elements of Jules Verne's romances—not only in their seeming improbability but also in the fact that a number of them will almost certainly come true. The Jordan Valley dam will change the economy of Israel. Solar power has had small-scale applications and may one day be a major source of world power. A tunnel beneath the English Channel is under discussion by the governments of England and France. Other projects, not so immediately feasible, must await new engineering techniques or a new political climate. Among them is a method of damming the Mediterranean, thus creating millions of acres of arable land.

ATLANTROPA—THE CHANGED MEDITERRANEAN

WILLY LEY

EVER SINCE the first long canal was dug, the first big bridge put up successfully, and the first tunnel finished, modern man has found it difficult to look at a map without considering whether something else could or should be done. To set the imagination working, a geographical feature on a map needs only to be narrow. There is that narrow piece of land separating the Red Sea from the Mediterranean Sea; result, the Suez Canal. There is the narrow strip of land tying North and South America together; result, the Panama Canal. There is in Greece the narrow Isthmus of Korinthos

(Corinth) separating the Gulf of Korinthos from the Gulf of Aigina; the Corinth Canal was the inescapable consequence.

The separating element may also be a narrow strip of water, say the East River between Manhattan and the tip of Long Island; the Brooklyn Bridge was built to span it. And if we return to the Mediterranean Sea, where narrow strips of land led to the Suez Canal and the Corinth Canal, we also find narrow strips of water around it, in a highly interesting and most challenging distribution.

At the eastern end of the Mediterranean Sea the small Sea of Marmara lies between the Aegean Sea, a northern tongue of the Mediterranean east of Greece, and the Black Sea. Both ends of the Sea of Marmara are decided narrows. At the western end there are the Dardanelles, a natural channel some 47 miles long, which has an average width of three to four miles but is less than a mile wide at one point. At the eastern—or, better, the northern—end of the Sea of Marmara is the strait which we call by the Greek name of Bosporus but which the Turks, who own it, call Istanbul Bogazi. It is 18 miles long, and almost three miles wide at the northern entrance. But it has a narrowest point with a width of only 2400 feet, with a strong current running from the Black Sea into the Sea of Marmara and on into the Mediterranean. At the southern end of the Istanbul Bogazi there is Istanbul on one side and Scutari and Kadikoi on the other side. The width at that point is a little less than a mile.

Decidedly this is a place which gives one the feeling that something ought to be done. In about 1875 one J. L. Haddan, who was then director of public works at Aleppo in northern Syria, proposed a tunnel; a bridge of the necessary length was impossible then. He even built a model of the strait and of his tunnel, but the Turkish government of the time would have nothing to do with it. The plan was rejected for political reasons, but this was not stated; instead, Haddan's plan was branded as a "wild and impossible dream." It was not; such a tunnel could have been built.

Much later (1949) an American engineer named Charles Andrew again suggested a tunnel under the Bosporus. The Turkish government remained noncommittal. It now looks as if they had already decided in favor of a bridge, because in July 1953, at the request of the Turkish government, the German firm of Stahlbau Rheinhausen submitted a plan and a bid for a bridge. The plan calls for a suspension bridge for automobiles and pedestrians (no railroad), with a length of 4550 feet.[1] But even if the bridge is started at once it will take five years to finish.

Moving westward in the Mediterranean area, we find another narrow strait near the middle of the sea. It is the Strait of Messina, with Sicily on one side and the Italian province of Reggio di Calabria on the other. About 50 years ago some not very definite plans for a tunnel were under discussion, but these have been finally abandoned recently in favor of a plan for bridging the Strait of Messina. A design was prepared in 1953 by the New York engineer Dr. David B. Steinman, at the request of the Italian Steel Institute. The bridge would not have to be much longer than existing bridges, but one of the difficulties is that, in the Strait of Messina, it is not only the waters which are turbulent. The ground is earthquake territory—one of the most destructive earthquakes of this century is actually labeled the Messina earthquake. And violent windstorms are almost the rule in winter. Taking all these conditions into account, Dr. Steinman proposed a specially designed suspension bridge, similar in general shape to the George Washington Bridge. The span between the two towers would have to be 5000 feet, some 800 feet longer than the span of the Golden Gate Bridge. The two side spans, between the towers and the land, would each have to be 2400 feet long or about the length of the side spans of the largest bridges now in existence.

The western end of the Mediterranean is even more in-

1. The George Washington Bridge in New York has a length of 3500 feet; the Golden Gate Bridge in San Francisco a length of 4200 feet.

triguing. Again we have the picture of two mighty land masses approaching each other until they almost touch. Almost, but not quite; they are separated along a 30-mile length by the Strait of Gibraltar. This opening, also characterized by a strong current flowing into the Mediterranean Sea, is considerably larger than the narrow straits at the eastern end. Its entrance, coming from the Atlantic Ocean, is more than 25 miles wide but narrows to about 7½ miles. What makes it so impressive is that it is flanked by two remarkable mountains, the Rock of Gibraltar on the north and Djebel Musa on the south. To the ancients these were the Pillars of Hercules. One version of the story has it that Hercules, having been sent to Spain to perform one of his superhuman labors, erected these two "pillars" to mark his travels; another version, a bit more specific, states that Hercules made these two pillars by cleaving a single mountain into two.

That latter version may even be true, provided you substitute the force of Nature for the strength of the son of Alcmene. It is certain that the Strait of Gibraltar did not exist in the far past, but nobody can tell with any certainty when it came into existence, or whether the break between the two continents occurred suddenly or was a slow sinking or disintegration that took many thousands of years.

Naturally (I am tempted to say) the Strait of Gibraltar has its tunnel plan, just like the English Channel and the Bosporus. The Gibraltar tunnel was worked out by a French engineer named Berlier. Because the strait happens to be especially deep at its narrowest point, the location chosen for the tunnel is from Tangier on the African side to a point west of Tarifa on the European side. The actual shore-to-shore distance along that line is 20 miles, but the tunnel would have to be 25 miles long. It would have to go rather deep, requiring long approaches, with the center section 1500 feet below sea level. Berlier's tunnel does not create a political problem, since both ends would be on Spanish territory. Whether a group of banks could be convinced that it is an economic necessity and would therefore prove profitable is a different question.

However, all these tunnel and bridge plans are comparatively minor projects in relation to the Mediterranean Sea, which because of its peculiar structure, shape, and location has really caused people to think.

It was in March 1928 that scientists and engineers first read, with intense amazement, about a plan outlining what could conceivably be done with the Mediterranean. The author was Herman Sörgel, whose Congo plan is described in Chapter 5. Nothing much was known about him at the time, except that he was an architect employed by the government of Bavaria. At first Sörgel called his scheme the "Panropa Plan"; later, presumably to avoid confusion with a similarly named political association aiming at European confederation, he changed the name to "Atlantropa Plan." The magnitude of Sörgel's proposal was such that some readers wondered whether he was actually serious or was engaged in a huge if obscure joke. To show that he was serious he published in 1929 a more comprehensive statement which was printed in parallel columns in four languages: English, French, German, and Italian.

The title of this publication was *Mittelmeer Senkung*, which means "Reduction of the Mediterranean." Sörgel began with the statement that the Mediterranean Sea in its present shape is a recent body of water. According to some geologists—there is no complete agreement among them on the figures though they are agreed on the facts—about 50,000 years ago the level of the Mediterranean was about 3000 feet lower than it is now. Since 44 percent of the area of the present sea is less than 3000 feet deep, the ancient sea was obviously much smaller. In fact, there was no Mediterranean Sea then, but only two large lakes, one of them east and one west of present-day Italy-Sicily. Europe was connected with Africa by three wide land bridges, one from present-day Spain to Morocco, one from Tunisia to Sicily and Italy, and one from Greece across the eastern end of the present sea. Later, when the glaciers of the Ice

Age melted away, a great deal of fruitful and possibly even inhabited land was drowned. If we want to, said Sörgel, we can get much of this drowned land back. All we have to do is to revoke the alleged feat of Hercules and plug the Strait of Gibraltar with an enormous dam.

It is an undisputed fact that the Mediterranean would shrink visibly if the Strait of Gibraltar were filled in, say, for argument's sake, by an earthquake. The Mediterranean is large, comprising a total of 970,000 square miles, just about ten times the area of the State of Wyoming. It is a warm sea, which means that much of its water is evaporating constantly. The evaporation losses are so high that the level of the sea would recede 5½ feet per year if the water were not replaced. The total mass of water which evaporates from the surface of the Mediterranean every year is a staggering figure—4144 cubic kilometers, which is 146,343,000,-000,000 cubic feet! [2]

Actually of course the level of the Mediterranean Sea does not recede because the evaporation losses are made up in various ways. Rain which falls into the sea replaces just about a quarter of the loss. Rivers also help, though not very much, since most of the rivers which empty into the Mediterranean are small and some of them are even seasonal. Only four of them can be called large, the Nile from Egypt, the Po from Italy, the Rhone from France, and the Ebro from Spain. The balance is maintained by water from the Black Sea and the Atlantic Ocean, hence the strong currents in the straits at both ends of the Mediterranean. The contribution of the Atlantic Ocean, which replaces two-thirds of the evaporation losses, results in a flow through the Strait of Gibraltar of 88,000 cubic meters (or 3,100,000 cubic feet) per second, which is twelve times the amount of the waters falling over Niagara Falls at high water!

In figures, the contributions of these four sources to the water budget of the Mediterranean are as follows:

2. A cubic kilometer is 1,000,000,000 cubic meters and each cubic meter equals 1.3079 cubic yards, or 35.314 cubic feet. A cubic yard equals 27 cubic feet.

	AMOUNT PER YEAR		
SOURCE	(cubic kilometers)	(millions of cubic yards)	PERCENTAGE
Atlantic Ocean	2762	3,612,000	66.65
rain	1000	1,308,000	24.11
rivers	230	301,000	5.56
Black Sea	152	199,000	3.68
Total	4144	5,420,000	100.00

This table is the basis of the Atlantropa Plan.

Obviously nothing can be done about the rain, and the rivers may as well be left alone. But it would be easy to build a dam across the Dardanelles. By itself this dam would not do any good because the Atlantic Ocean would simply provide an additional 200,000 million cubic yards per year. Still, Sörgel proposed to start at that end, largely because there would be no water-level difference on that dam during construction.

The big job would be the Gibraltar dam. Like Berlier's proposed tunnel, the dam is not planned to cross the strait at its narrowest because the depth there is more than 1600 feet. It would have the shape of a horseshoe, with its open end toward the east. Along the proposed curve of the dam there are many shallow places and the deepest point is about 1000 feet below sea level. The length of the dam would be 18 miles; its crown would be 165 feet wide but its foundation would have to have ten times that width to withstand the water pressure that would soon develop. The estimate of a drop of 5½ feet per year in the level of the Mediterranean is based on the assumption that the sea would receive no water from any source. Since the rain and the rivers still provide some, the actual recession of the level would be about 40 inches per year. But this would cause a drop of 33 feet within ten years after the completion of the dam, and this is enough of a difference in level to be utilized to produce electric current—and, considering the amount of water which could be sent through the turbines, to produce it in quantity.

In drawing up the plan, Sörgel did not think in terms of years, and only rarely in terms of decades, but often in terms of centuries. One century after the completion of the Gibraltar dam the level of the Mediterranean Sea would have gone down 330 feet, and by that time a total of 90,000 square miles of new land would have appeared above the surface, with the gains almost proportional for most of the countries bordering the sea. Spain's largest gains would be in the area of the mouth of the Ebro River, France's largest gains in the area of the mouth of the Rhone River. The two islands Mallorca and Minorca would have become one, as would Corsica and Sardinia. Italy would have gained on both coasts, and most of the northern end of the Adriatic Sea would have become the Adriatic Land. Sicily would have grown enormously, and Tunisia too; they would almost but not quite touch. A strait would also still remain between Italy and Sicily but it would be a very narrow one.

From that point on, a further reduction of the sea's level would have little effect on the western part of the Mediterranean but much more on the eastern part. So Sörgel proposed as the second step, a hundred years after completion of the Gibraltar dam, two more dams; one across the remaining narrow strait between Sicily and Italy, say between Messina on the island to a point north of Reggio di Calabria on the toe of the Italian boot; the second to bridge the gap still remaining between Sicily and Tunisia.

After these dams were completed, the western half of the sea would be stabilized by permitting enough water to enter from the Atlantic to maintain that level. The eastern half would be permitted, in the course of another century, to sink for an additional 330 feet, which it would do all the more readily since virtually its only source of supply would be the Nile. Then the eastern half would also be stabilized, partly with water from the Sea of Marmara and hence from the Black Sea, partly with water from the western half. When completely adjusted in accordance with this plan, the Mediterranean Sea would hold 350,000 cubic kilometers less water than it does now. This amount of water would be distributed over all other oceans, since they are all interconnected, and would raise the sea level everywhere else by three feet.

In the Mediterranean area the final result would be 220,000 square miles of new land and hydroelectric power plants of virtually unlimited capacity in a number of places well distributed over the area: at least two in the Gibraltar dam, one each at the mouths of the Ebro, Rhone, Po, and Nile, at least one in the Dardanelles dam, and a minimum of two each in the two dams separating the western half of the sea from the eastern half. There would certainly be no lack of power for anything the inhabitants of the area might wish to do.

Sörgel considered that the final goal of the Atlantropa Plan was the fusion of the European and African continents; he entitled a later book (published in 1938) *The Three Big A's*—America, Asia, and Atlantropa. But he also repeatedly pointed out that his plan could be stopped at any moment after the level of the sea had dropped, say, 50 feet, and still realize its chief purpose. Such a comparatively small reduction of the sea's level might not provide much new land, or any land of value, but it would create the means of producing enormous amounts of electric power.

If Sörgel's main idea—the dams across the Dardanelles and the Strait of Gibraltar—is ever carried out it is highly probable that the men entrusted with the Atlantropa project will stop when the sea level has fallen 50 or 75 feet.

It is important to remember that none of this must interfere with shipping. Damming the Dardanelles might not be hard, but ships still would have to get from the Mediterranean to the Sea of Marmara and the other way round. For this, canals would have to be built, with a series of lock gates to take care of the difference in level. The Suez Canal would have to be lengthened and locks built at the Mediterranean end, and, most important, canals and locks would be required for the connection between the "inner sea" and the Atlantic Ocean.

The Gibraltar locks especially would have to be large enough to accommodate the largest ocean liners, battleships, and aircraft carriers. Every additional 50 feet of level difference to be overcome means one more lock in each canal and adds considerably to the length of the canal. The biggest gain would actually be made by the first 50 feet of level drop. This drop would provide power without doing too much harm to existing installations.

It would also accomplish something else which has not yet been mentioned. The figure of 3,100,000 cubic feet per second for the inflow through the Strait of Gibraltar represents the difference between two flows which take place there. At the surface the influx of water from the Atlantic is greater than that figure, but at the bottom there is, according to Sörgel, a flow in the opposite direction which brings cold bottom water from the Mediterranean into the open sea. This outflow is accused of forming a cold-water cushion outside western Europe which deflects the Gulf Stream into the northern Atlantic. Presumably, if this cold-water cushion were absent, the Gulf Stream would flow into the English Channel and thereby warm northern Europe more efficiently. A dam across the Strait, stopping any flow in either direction, would take care of that.

The reason many European experts who have studied the Atlantropa Plan, and many who are very much in favor of it, would like to stop after 12 or 15 years is that the continuation would prove costly. If the Mediterranean level were lowered 60 to 75 feet, about the only extra expenses required would be for equipping the Suez Canal with a lock somewhere near Port Said and deepening the Corinth Canal. But once the recession of the level goes beyond 100 feet, every single harbor along the shores of the Mediterranean stops being a harbor. Barcelona, Marseille, Genoa, Naples, Taranto, Trieste, Fiume, Haifa, Tel Aviv, Jaffa, Alexandria, Bengasi, Tunis, Bizerte, Algiers, Beirut, and Oran—to name only the larger ones that come to mind at once—would all be miles from the sea. Some of them might be "salvaged" by means of canals to the new shoreline, but many would become completely worthless, at least as harbors. Obviously, beyond a certain point the plan runs into the law of diminishing returns.

Another serious argument for stopping the Atlantropa

Plan at an early stage is the volcanism of the Mediterranean area. Even now conditions are not as stable there as one would wish and it is feared that removal of the weight of water from these unstable areas may lead to earthquakes and volcanic eruptions. This, of course, is a field in which nobody can claim to have experience. It is possible that such fears are unfounded, but it is by no means possible for anybody to claim that they *are* unfounded. However, it seems logical to assume that the danger would increase roughly in proportion to the weight of water removed and that the results of a minor reduction in sea level would be minor too.

Of course under present conditions the Atlantropa Plan is politically impossible. It would need the cooperation of more than a dozen different nations. Some, such as Spain and Italy, would gain relatively much, while others would actually lose something; England, for example, would lose control of the Strait of Gibraltar. It is therefore unnecessary to discuss the engineering difficulties involved in the construction of a dam of such magnitude as the Gibraltar dam. The political situation of today makes the engineering problems future problems, and we can't tell how an engineer 50 years from now would go about solving them.

There can be no Atlantropa until a united Europe is a reality and until the control of the Strait of Gibraltar has developed into a purely commercial problem, with all military aspects absolutely missing.

Interestingly enough, the situation which prevails for the whole Mediterranean also holds true for a much smaller body of water in its immediate neighborhood—the Red Sea, part of the Great Rift Valley.

The Red Sea is, in round figures, 1200 miles long, and on the average less than 200 miles wide, except in its southern part where a maximum width of 250 miles occurs. Its surface has an area of about 180,000 square miles. The land bounding it along both shores is desert. Its northern end is closed except for the one artificial opening of the Suez Canal. Not a single river empties into the Red Sea.

Its southern end is the sharply constricted Strait of Bab el Mandeb, which is further narrowed by the existence of the British-owned island of Perim, which divides the strait into two channels. The one to the east of the island is two miles wide, the one to the west 16 miles.

Since the Red Sea is a part of the Great Rift Valley it is much deeper than one would expect. The average depth is 1600 feet, but a number of extra-deep depressions have been measured, one of 4200 feet and one even of 7200 feet.

The "Red Sea Plan," worked out by the French engineer René Bigarre and published in 1940, shows many similarities to the Atlantropa Plan, with some features of Dead Sea development projects thrown in. It simply consists of a dam across the Strait of Bab el Mandeb with Perim as an off-center anchor. Naturally there would have to be a canal with lock gates connecting the Gulf of Aden with the Red Sea, and the Suez Canal would also need a lock gate at the Red Sea end, in the vicinity of Suez. After these had been built evaporation could take its course. It is so enormous there that Matthew F. Maury, a United States naval officer whom many call the Father of Oceanography, estimated a century ago that the level of the Red Sea would sink 23 feet a year if no water could come in from the Gulf of Aden and the Indian Ocean.

René Bigarre thinks Maury's estimate too high and counts on only 0.4 inches per day or slightly more than 12 feet per year. Even this is enormous, amounting to a removal of 153,200 million cubic feet per 24-hour period. If Bigarre's figures are right, power generation could start about five years after completion of the dam; if Maury's should be correct, it could start in two years and a few months. If all the water then required to maintain the Red Sea at its new level were sent through turbines, the power output would be about 240 million kilowatt-hours daily. To produce that much by means of steam turbogenerators you would have to burn 200,000 tons of coal daily!

Some of the present Red Sea area would become dry land in the process, the area depending of course on the total level drop permitted. If the drop in level should be 100 feet, the dry areas would comprise 10 percent of the present total area of the Red Sea, possibly 12 percent. As land this land would be useless; as a source for salt it would be useful. The Red Sea is saltier than the open ocean right now and evaporation would make it saltier still. The yield would run into millions of tons.

From almost any angle the Red Sea Plan looks like a miniature version of the Atlantropa Plan. Carrying out Bigarre's plan would virtually provide a testing ground for Sörgel's ideas. It certainly should come first. The Bab el Mandeb dam is certainly an engineering possibility. The political difficulties are not so great that they might not be overcome. But nobody in the general area of the Strait of Bab el Mandeb needs such quantities of electric current.

The conquest of space offers perhaps the greatest engineering challenge of modern times. During the next century it may conceivably change man's physical and ideological environment as profoundly as all the discoveries which have preceded it. In 1962, members of the Twentieth Century Assembly convened at the Harriman Campus of Columbia University to discuss papers written by eight authorities "to render intelligible the complex issues and alternatives which confront us on the fantastic frontier beyond the earth." The subjects included the economic, governmental, military, and judicial problems of the space revolution. One of the outstanding papers, here reprinted, discussed the technical prospects. It is written by the Professor of Aeronautics and Astronautics at the Massachusetts Institute of Technology, formerly Chief Scientist of the United States Air Force, who in 1960 was elected President of the Institute of Aerospace Scientists.

OUTER SPACE: THE TECHNICAL PROSPECTS

H. GUYFORD STEVER

A SERIOUS PROGNOSIS of the technical prospects of space flight requires more than a mercurial judgment about the quick attainment of some of the projects now being discussed. It requires an appreciation of the history of technology, of how new technologies unfold. There is a striking parallel between the history of the airplane and the history of space flight to date. A review of this parallel can show the kinds of indicators to be looked for in estimating the prospects for the future of space flight.

The Analogy of the Airplane

Man's early dreams of flight in the atmosphere like a bird were intermingled with his dreams of space flight to other planets throughout the solar system. Confusion existed because natural philosophers did not have an accurate picture of the extent of the atmosphere. The early myths of manned atmospheric and space flight include that of Daedalus, builder of the Minoan Labyrinth, and his son Icarus, that of King Bladud, tenth legendary King of England and father of King Lear, and a host of other tales, some of which may well be based on early, probably tragic, experiments in flying with bird-like wing constructions. As science began to develop in the Middle Ages, more practical dreamers, if we may use such a term, such as the Franciscan monk, Roger Bacon, and later Leonardo da Vinci, had ideas which might have led to practical embodiment had the technology been sufficiently advanced. For over two hundred years before flight was achieved, physical experiment and attempts to fly increased steadily, and more and more men began to get the concept of powered flight.

The engineering basis of flight was laid in the early nineteenth century, long before the Wright Brothers first flew, by Sir George Cayley, a British minor nobleman, who was a brilliant engineer in many different fields. His accomplishments in aeronautics, though not widely appreciated, were astounding. He was the first to realize that the airplane would attain the lift needed to counteract its weight by a thrusting device including a propeller and an engine which would overcome the drag of the air. Among his many other aeronautical accomplishments he designed and built the first model of a practical airplane. But, more important, he had a very clear vision of the future and in 1809, almost a century before manned controllable powered flight was achieved, he wrote:

> I may be expediting the attainment of an object that will in time be found of great importance to mankind, so much so that a new era in society will commence from the moment that aerial navigation is familiarly realized—I feel perfectly confident, however, that this noble art will soon be brought home to man's convenience and that we shall be able to transport ourselves and our families and their goods and chattles more securely by air than by water and with velocities of from 20 to 100 m.p.h.

Though his numbers fell short of the mark, he had the spirit of the modern development of air transportation. These were the words of an imaginative but still practical engineer.

On the other hand, few basic research scientists had anything to do with the attainment of flight, nor were they sanguine about its use. They generally ignored the field or discounted it. For example, Lord Kelvin, one of the world's great research physicists, said in 1896, only seven years before the attainment of controllable manned powered flight, "I have not the smallest molecule of faith in aerial navigation other than ballooning."

Even after the achievement of controllable powered, manned flight, and after many people had followed the Wright Brothers' lead, it was still difficult to foresee the future. In 1908 the Wright Brothers, who were still leading in the development of airplanes all over the world, delivered to the United States Army an airplane to fulfill a contract which called for a flight speed of about 32 m.p.h. It was constructed of airplane cloth and a hickory wood frame; it had two small nine-foot propellers geared by belt drives to a single motor of which the power output was about 25 horsepower, lower than that of practically any modern automobile. You recall the pictures of the Wright Brothers' Flyer, with its fixed horizontal tails in the front, vertical rudders to the rear. It did not even have wheels—just skids. It was normally launched with a catapult mechanism, though occasionally Wilbur Wright was skillful enough to take it off on wet grass. A standard stunt in those days was for a man or two to push on the rear of the wings to help the airplane get started.

Still the concept of flight was exciting to enough people so that in its infancy many predictions were made of the technical prospects of the airplane and of its use to mankind. Not all were imaginative. In 1910, for example, the British Secretary of State for War said, "We do not consider that airplanes will be of any possible use for war purposes."

The first uses of the airplane which spurred its development were military. It was an improved means of performing certain limited military tasks such as observing the enemy. To most minds its function was to replace the cavalry as the eyes of the army and the balloon as the spotter for the artillery.

It was very difficult for a practical man in 1909, when the first Wright planes were being adopted for military use, to conceive of commercial air transportation. Although calculations of the air-transportation economics of those days vary quite considerably, they point up some basic facts. In 1909 a plane could travel at 42½ m.p.h. with a pilot and a single passenger. The plane had a useful life of only a small number of hours—possibly 30—and it cost $30,000. The cost per passenger mile might then turn out to be something like $25 per passenger mile or, in 1960 dollars, $80 per passenger mile. Today the operating cost of a jet airplane which flies more than ten times as fast and has a useful range of almost

100 times as great with 100 times as many passengers is only a few cents per passenger mile.

No one accurately foresaw the shape of things to come for the airplane. Those who had faith that technology has a future came closest to predicting the future. There is a story, possibly apocryphal, that the head of the Astor business enterprises said that important men would conduct their business by traveling in airplanes in 50 years. He did not worry about the limits to the load-carrying capacity of wood and fabric airplanes. He did not worry that there were limits in the power available. He did not worry that flying an airplane at that stage was dangerous. He did not even stop to consider the tremendous development cost.

Astor went right to a useful purpose that the knowledge of the day promised, and his faith in technology proved right. The structures changed from cloth and wood to metal. Steel and the light aluminum and magnesium alloys were developed, and the technique of stressing the skin instead of using a bracing framework brought aeronautics to its modern era. The engines developed from a few tens to many thousands of horsepower; internal combustion gasoline engines with propellers were replaced by turbojet engines. The vast technological improvements in every field of engineering associated with the airplane have made commercial flight commonplace.

The parallel between the story of the airplane and that of the applications of our space technology is obvious. In the military context, the first space concepts were observation satellites. The achievement of a bombardment capability from space and space combat is now being given serious thought and development. Eventually military operations may well be conducted simply for control of space as in the past they have been conducted for control of the air. In the context of peaceful applications, there has been some slower development.

Only after major emphasis on military uses do we now appear to have within our reach world-wide communications by satellite relay stations, a world-wide weather observation and prediction service using satellites, and a world-wide navigation system for ships using navigation satellites.

The Basis for Predictions

The lessons in prognosticating the technical possibilities of space flight that can be learned from this brief consideration of another great technology are numerous. For example, most people, over the long run, fall short of the mark in their predictions. Developments follow the lines of practical use. Military developments lead the way to nonmilitary applications.

I have history in mind, then, as I attempt here to look ahead to the technical prospects for outer space. Moreover, I have in mind certain very present factors which bear on any forecast.

In the latter half of the nineteenth century, when classical science was flourishing, J. Henri Poincaré wrote in *La Science et l'Hypothèse*: "For a superficial observer, scientific truth is beyond the reaches of doubt; scientific logic is infallible and, if scientists sometimes err, it is because they have misunderstood the rules."[1] If the task of presenting the technical prospects for outer space depended only upon understanding scientific truths, the future could be plotted with reasonable simplicity and confidence for some time ahead. But progress in space will not be essentially or solely scientific. It will involve engineering, in which the laws of science play an important but only partial role. Thus progress in space, like that in all engineering projects, will be critically affected by economic and social factors.

For decades space progress can be made by practicing in new and generally more expensive embodiments the arts we already know. It will depend upon engineers who must improve the design of existing equipment, design similar equipment in larger sizes, and develop new devices in fields of engineering where the principles are well known. We can already identify some of the areas in which those devel-

1 Translated from the French.

opments will be made. Steady but not overwhelming gains can be made in liquid and solid propellant rocketry. Nuclear rocketry and electrical particle rocketry are being developed with the promise of vast improvement in space capability. Some of the most important but least publicized gains in the recent past and expected gains in the near future are in the fields of structural design and materials. Auxiliary power is a key field of future development. Communications, radio and inertial guidance, and other space navigation developments will be needed before useful space accomplishments can unfold in large number. The engineering of life-support equipment for human flight is a relatively new field which offers great promise of improvement.

It is clear also that space progress will depend upon the financial support given to the development organizations of which we already have many more in this country than we are using efficiently. Moreover, it will depend upon the size of the continuing military effort.

However, any prediction based solely on our current technology, with reasonable estimates of government interest and financial support, would most certainly lead to an underestimate of the technical prospects for outer space. As any student is aware, future progress in engineering will depend upon developments not known in today's art; and the talented young men and women who are now going through training in engineering and science, much better equipped than earlier generations both in background and in their approach to education, will march to the future of technology more rapidly than we now estimate. One can be sure that there will be major new developments which are not foreseen today, and that some of them will move the space program forward faster and farther than we can predict.

At the outset of this chapter, then, I want to declare that I am an enthusiast for the long-term potential of space flight. I can best describe my attitude by telling an anecdote about a foreign visitor who took a taxi tour of our national capital. When shown the Archives building, on which there is inscribed a quotation from Shakespeare's *The Tempest* which reads "What is past is prologue," the foreign visitor was a little puzzled, for he did not have a good command of the English language. He asked his taxi driver if he knew what the saying meant. The taxi driver answered, "Sure, bud, that means you ain't seen nuthin' yet."

I believe that we have only scratched the surface of the technology of space flight. I believe that we have the trained engineers in the aerospace field with enthusiasm and vision who can achieve their promises.

Space Flight Velocity Requirements

The velocity requirements for various space missions have nothing to do with the past, present, or future state of technology. They come out of a very old branch of science, celestial mechanics, which began with the ancients as they studied the motion of the stars, was given a big boost by Copernicus and a mathematical foundation by Kepler and Newton, and grew to its peak many decades ago as astronomers made the system accurate. In fact, it was a science which was almost in mothballs until the new-found rocket technology returned it to prominence.

The Problem of Propulsion

The key technology in space flight is propulsion. Rocket boosters are now capable of accelerating useful payloads to the very high velocities which are required to orbit the Earth, to escape the Earth and go to the Moon and the other planets, and to orbit the Sun. Most alert readers of the newspapers in recent years have amassed a few characteristic numbers which describe the high velocities required for space flight. For the purposes of this chapter, in describing the speeds, it is worth introducing an illustration (see graph: *Velocity Requirements*, etc.). Incidentally, this graph could have been prepared by Sir Isaac Newton using his newly enunciated Law of Universal Gravitation and the mathematical tools available to him.

The graph shows the velocity required for a body to move

on an elliptical orbit starting and terminating on the surface of the Earth as the ballistic missile does, in a circular satellite orbit around the Earth as a Sputnik, a typical Earth satellite, or as the Moon does, and in an elliptical orbit changing from the Earth's orbit around the Sun to one of the planet's orbits around the Sun. Though Sir Isaac would have been capable of plotting such a graph, he certainly would have objected to the entire proceeding as not being scientific, since only a dreamer would think that man could ever have the capability of attaining such velocities in any useful vehicle.

Velocities are given in feet-per-second. Many readers are more familiar with velocities given in miles-per-hour. It is easy to convert approximately from feet-per-second to miles-per-hour by taking two-thirds of the number. For example, 15,000 ft/sec is roughly 10,000 m.p.h.

First, consider the speeds required for ballistic missiles. The very high speed of 5,000 ft/sec, about one mile/sec or 3,600 m.p.h., enables a ballistic missile to achieve a range somewhere between 200 and 250 miles. An increase in speed to 15,000 ft/sec, on the other hand, enables the vehicle to attain a 1500-mile range. A speed of less than double that —between 23- and 24,000 ft/sec—permits a factor of four increase in range from 1500 to 6000 miles. An increase in speed from about 23,500 to 25,000 ft/sec increases the range from 6000 to 12,000 miles; and a very slight increase of velocity over the 25,000 ft/sec puts a satellite into orbit at low altitude.

Further increases in velocity capability from 26,000 feet to about 36,000 ft/sec permit the satellite orbit to be established at increasingly high altitudes to a point where at something over 36,000 ft/sec the gravitational pull of the Earth can be entirely escaped so that the vehicle would then be in orbit around the Sun just as the Earth is. With a few thousand ft/sec more in speed, a space vehicle can get to the regions of Mars and Venus, say at 41,000 ft/sec, to Mercury with 45,000 ft/sec, Jupiter with 51,000 ft/sec, and so on.

These speed requirements are well known to space engineers; in fact all of them have these numbers at the tip of their fingers at all times and, in this era of advertising pub-

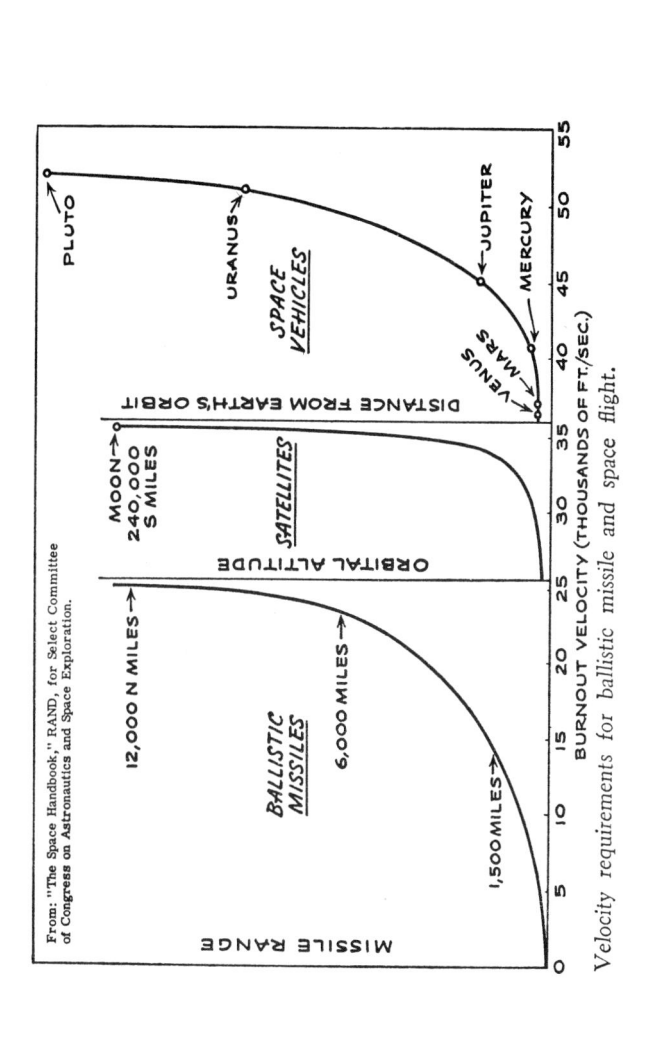

Velocity requirements for ballistic missile and space flight.

licity and public speeches, they are not only at the tip of the space engineer's fingers but also at the tip of his tongue.

The speeds given on the chart are minimum to achieve the objectives. If the mission requires some special maneuvering such as landing on a planet, the speeds are somewhat higher. For example, if one is considering a round-trip lunar flight in which the vehicle takes off from the Earth, uses a rocket to brake its speed as it decelerates to land safely on the Moon, takes off from the Moon, and comes back to the Earth using atmospheric braking here on the Earth, the speed capability of the rocket should not be just 36,000 ft/sec, but something like 60,000 ft/sec. Likewise, if a spaceship is required to go from Earth to Mars on the minimum velocity of about 37,000 ft/sec, the spaceship can do this only when Mars is in the optimum position and it could not use any rocket braking or other maneuvers around Mars. On the other hand, with a capability of 100,000 ft/sec change due to rocket thrust, instead of just the minimum 37,000 ft/sec speed capability, it could propel itself there in 15 or 20 days; if it had 300,000 ft/sec it could make the trip in 10 days. So the figures given on the graph are misleading with respect to space missions of an advanced nature. In reality one would like to be able to design very high-speed increments into the rocket boosters which enable the space vehicle to perform its mission.

Some Comparisons

One might digress here in order to put these very large velocities into context with other high-velocity devices that are well known. The long history of ballistics and firearms has led to developments in which small-arms can now have velocities from 1000 ft/sec to between 2,000 and 3,000 ft/sec. Certain very high-performance guns can go to higher velocities, and in the laboratories for special research purposes there are gun-type devices which go to 10,000 ft/sec and more. Jules Verne in describing his imaginary trip to the Moon employed a very long gun barrel with a very special new explosive to propel his ship to the Moon. Even if one could use some kind of gun-like projector for a ship for space flight, it would have several drawbacks. The first of these is that, since the highest velocity is attained right at the end of the gun barrel, which presumably would be within the atmosphere, all the difficulties of high-velocity frictional heating would plague the vehicle during takeoff. Moreover, the velocity loss in the atmosphere due to drag would be very large. In addition, there would be immense problems of high acceleration loading (high "G's") on the vehicle because the full acceleration would take place in the very short gun barrel. For these reasons the rocket principle is used.

Rockets have the tremendous advantage that the accelerations are least in the beginning and stay relatively small, small enough to be withstood by humans and by delicate instruments. The very high velocities can be reached because the accelerations occur over long periods. Furthermore, the extreme velocities are not reached until the denser portions of the atmosphere are cleared by the vehicle.

Rocket Booster Technology

The concept of using rockets to attain the very high velocities for space flight is rather old. In fact, it would be difficult to pinpoint accurately the first man to conceive this. In the nineteenth century a Russian, Tsiolkowsky, a minor schoolteacher, discussed rocket power as the means by which the high velocities required for space flight could be reached. A German named Ganschwindt independently did the same.

Robert Goddard, an American physicist who started thinking along these lines during World War I, also discussed and placed on a much more scientific basis the calculations for rocket propulsion needed for space flights. He spent his whole professional career in efforts which initiated modern liquid propellant rocket technology and designed vehicles which were the forerunners of today's space vehicles, reaching in 1926 the point at which his first propellant rocket vehicle was fired.

Most of the achievements in space flight have been made using the liquid propellant rockets which were pioneered by Dr. Goddard, developed to a reasonably high state of the art by Germans in their research and development leading to the V-2 and other rocket weapons used in World War II, and developed further in both Russia and the United States mostly for ballistic missiles and only lately for spacecraft. The modern interest in rocketry revived a much more ancient type, solid propellant rocketry, started by the Chinese in the twelfth century and used sporadically but relatively ineffectively in warfare from that time until World War II, when a large number of rockets using solid propellants, such as anti-tank air-to-ground rockets, anti-submarine rockets, and bazooka rockets for infantry against tanks, became quite effective weapons. In the period following World War II solid propellant rockets also have been developed to a point where they are now figuring in current and future space plans. Though their performance is not yet quite up to that of the liquid propellant rockets, this is partially compensated for by their higher reliability and greater simplicity in operation.

Where do we stand with respect to the speed increment that can be given to a vehicle as it is shot off into space? Not only can we fire ballistic missiles more than a quarter of the way around the world; we have established circular satellites around the Earth and sent vehicles toward the Moon. Beyond the Moon, vehicles have escaped the gravitational pull of the Earth to pass near Venus, be captured in the gravitational pull of the Sun, and remain permanent satellites of the Sun. According to the chart this means that we have attained velocities in the region of 40,000 ft/sec. This represents quite an advance in speed capability when we recall that in World War II, when the V-2 was put into operation, the best speed was a little more than 5,000 ft/sec and that, only 34 years ago, Goddard's rocket got only to 184 feet in altitude.

Too often, in considering the advances made in space boosters over the recent decades, exaggerated emphasis is laid on the rocket engine. The improvement of the performance of the rocket engine is only part of the story. An important part has to do with the improvement in vehicle design in which the relative weight of the vehicle components has constantly been reduced; and there is also the final element of design to obtain the very high velocities desired, that is, multistaging. All very high-velocity space vehicles and even ballistic missiles—the longer range ones—are boosted by multiple stage rockets. The principle of this staging is very simple: if a single stage rocket can, say, boost a payload to half of the velocity required for a given mission, then the full velocity can be achieved by adding a larger booster stage. This booster stands in weight ratio in the same relationship to the original rocket plus payload as does the original rocket—which now becomes the second stage—in relation to the payload. This staging device has the advantage of enabling the high speeds required for the mission to be obtained; it has the tremendous disadvantage that the multiplying factor mentioned goes up very rapidly.

Suppose, for example, a mission of 25,000 ft/sec is considered. If a rocket can be designed to push a payload of 1,000 pounds to 12,500 ft/sec with the total rocket weight being ten times the weight of the payload, or 10,000 pounds, then the full velocity for the mission—the 25,000 ft/sec—can be achieved by taking the 10,000 pounds of the first rocket, and with the same ratio of ten times for a larger booster stage, or 100,000 pounds, the 100-pound payload can be boosted to 25,000 ft/sec. Carrying this same reasoning a little further, if the mission calls for 37,500 ft/sec which would permit it to escape the gravitational pull of the Earth, the 100,000 pound total vehicle would again have to have a still larger booster stage added which was ten times its weight—or a million pounds. So the staging principle allows one-tenth by weight of the first stage rocket to be boosted to 12,500 ft/sec or one one-hundredth by weight of a two-stage rocket to twice that speed or 25,000 ft/sec, or one one-thousandth by weight of a three-stage rocket to a speed of 37,500 ft/sec. One can carry on the arithmetic from there and see that it gets both expensive and discouraging to increase the stages far beyond three or four.

Since the days of the V-2, when the single-stage velocity was of the order of 5,000 ft/sec, improvements in rocket efficiency and in structural efficiency have made it possible for a single stage to reach 15 to 20,000 ft/sec. No single-stage rocket has yet reached the 25,000 ft/sec required for orbiting the Earth, though vehicles which are almost single-stage vehicles—like the Atlas, which instead of dropping off the stage only drops off some excess rocket engines, and is therefore called a one-and-one-half stage vehicle—have reached this velocity of orbiting. With today's technology one thinks of one or two stages for long-range ballistic missiles, two or three stages for orbiting vehicles, and three, four, five, or six for vehicles to go to the Moon and to escape the Earth's gravitational pull—to go to Venus or Mars or just to become ordinary satellites of the Sun. Research vehicles have been used with as many as seven stages.

One may ask the question: Can there be a radical increase in the speed increment which is obtainable from a single stage of a booster? As indicated before, such an increase must come from improving the efficiency of the engine itself or from the improvement in the structural efficiency.

Let us first look at the efficiency of the rocket itself. Over the past period of development the rocket motor design has been given a tremendous amount of attention, but for a given rocket propellant such as liquid oxygen and kerosene the expected improvement in rocket motor design cannot be very great. A given engine may be made somewhat more efficient with long and expensive development programs on the turbine fuel pumps, on the inlet design, on the jacket cooling, on the materials used, and so on. But only small gains can be made. Larger gains can be made by changing rocket propellants completely, and steps have been taken along this line. The standard rocket propellants were liquid oxygen and kerosene for the very long-range ballistic missiles and spacecraft of the recent past. More recently liquid oxygen-liquid hydrogen engines have been developed with higher performance figures. There are other possible improvements using liquid hydrogen-liquid fluorine and so on. In the solid propellant field there are also possible new propellant combinations that can be made, and such tricks as making a combined liquid and solid propellant rocket are under development. One can expect some improvements then, in the propellant efficiency, but they will come in small increments and only following long-term and very expensive projects.

Considering the possibility of improving the velocity increment obtainable by a single stage by increased structural efficiency, one should point out that from the days of the German V-2 only about 70 percent of the booster was rocket fuel. Today engineers have been achieving almost 90 percent. Any small increment at this high percentage is valuable, but even a small increment is extremely difficult to obtain. Thus one can expect some improvements along this line, but nothing radical, barring of course one of those unforeseen inventions that cannot be taken into account in this prognosis.

Costs of Boosters

The cost of boosting a payload to the high speed required for its mission—including the development of boosters, establishment of complex launching bases and operating them, and the manufacture of the hardware and the fuel—represents a major share of the total cost of the space program. In the first place, the development costs are huge; the boosters are complicated technological devices which require large design, development, and test teams to get them into any reasonable state of operational readiness. In the long run, however, development costs become less important than operational costs.

One of the major operational costs of space boosters is the fuel. Every vehicle, be it a long-range ballistic missile or an orbiting vehicle, takes off loaded as high as 90 percent of its total weight with fuel which is burned in the mission. When one considers that these vehicles range up to 200,000 pounds now, and will range to millions of pounds in the future, one realizes that the fuel cost alone will be considerable. Rocket booster engineers know this full well; and in their search for high-performance fuel combustion for their liquid and solid propellant rockets they also keep an eye on the production cost figures of the propellants.

However, one should not be too discouraged by the fact that such a large percentage of the take-off weight is fuel. We already have experience with operations in which very large amounts of fuel are used but which have become economically feasible—for example, one of the standard jet aircraft used today by commercial airlines. With a take-off weight of about 280,000 pounds, the fuel weight of such a plane is 122,000 pounds, or between 40 and 45 percent of the take-off weight. For a payload of the order of 36,000 pounds, between a third and a quarter of the fuel weight is expended in a flight. If one considers not the typical passenger jet airliner but the long-range bombers designed for a maximum fuel capacity in order to achieve a maximum range, one finds that instead of between 40 and 45 percent of fuel in the take-off weight, it runs to 50 to 60 percent. So a mission which involves expending most of the initial weight of the vehicle in fuel consumption is not necessarily something that cannot be made economically feasible and even profitable.

In current space operations one of the greatest expenses arises from the fact that the booster vehicle is used for only one flight. R. C. Truax, Director of Advanced Development at Aerojet General's Liquid Rocket Plant, told a panel of space writers in New York that "if an airliner today were to be used only once on a cross-country trip and then thrown away, the fare per passenger just to pay for the airplane would be around $30,000." Clearly the practice of using a rocket booster for a space mission only once must be changed, an objective on which efforts are now under way.

Before describing some of the techniques of recovering the booster vehicles for space missions, it is of interest to establish in the reader's mind a cost figure for space operations to be used as a standard. It is a difficult figure to calculate accurately because a typical space mission involves not only the cost of the hardware and the fuel as purchased from the manufacturers but also the cost of the launching team and the team that tracks the vehicle in its flight, and so on. Since the organizations which carry out these functions are complex, it is somewhat difficult for a cost analyst to track through government organization and make a fair assessment to each of the many organizations involved for the cost of their share of the operation. Even recognizing this difficulty, engineers today use as a standard the cost of putting a single pound of payload into a circular orbit about the Earth at an altitude of about 300 miles. Rough cost estimates for various booster systems and various kinds of projects show that the costs run from $1,000 to several thousand dollars per pound of payload in orbit.

The long-range objective of booster engineers and designers is to cut this cost per payload pound in orbit by a factor of at least one-tenth and possibly by one-hundredth. It is not easy to achieve, at least by using the techniques at hand, but there are some hopes. A reduction in cost by the order of one-tenth might almost be achieved by making completely recoverable booster systems; there are many such proposals now under consideration. One generic type employs as the first-stage booster a kind of flying vehicle which, after taking off vertically and boosting the upper stages to some reasonable velocity, possibly a few thousand feet-per-second up to 10,000 or 12,000 ft/sec, converts itself into a flying vehicle which is flown manned or unmanned and landed as a conventional high-speed aircraft. Some designers would prefer to see the engines of this first-stage booster of the same type as current high-speed airplanes—namely, turbojet engines—reasoning that such engines give added convenience, reliability, and low fuel consumption. There are other proposals to use recoverable schemes involving parachutes and recovery systems such as snatching the returning launching vehicle in the air by a large helicopter. Though at first these sound complex and unreliable, more detailed examination indicates that they have some reasonable degree of feasibility.

Whatever the recovery system that is developed, one can be sure of two things: that such a system is entirely feasible, but that the actual development costs will be very large. It is the kind of complex development that requires few new basic scientific principles—only the application of a large amount of engineering design and effort. In the mind of the author, the development of such a system of the recovery of the early first stages in boosters is inevitable.

The cost per launch varies widely since different missions have different payloads. A few of the scientific missions of the past have used very small vehicles. For example, our first Explorer weighed only about 30 pounds, and our first Vanguard only three to four pounds. The first Mercury manned orbital capsule will weigh 2,000 to 3,000 pounds. Eventually tens of thousands of pounds of space vehicles will be sent into orbit.

For the small number of launch types of scientific and exploration missions the simple expendable booster systems will pay off best because their development costs are much less than those of recoverable systems. The many-sectioned solid propellant boosters will be somewhat lower in cost per launch than the liquid propellant systems, though as the number of launches increases, the liquid propellant systems approach the solids in cost. Also, as the number of launches go up, the cost of expendable systems goes down, but rather slowly. For missions where the number of launches begin to grow it soon becomes desirable to consider recoverable systems as discussed above—either those fully or even partially recoverable. The development cost for fully recoverable systems will be large, but not substantially more than for partially recoverable systems. With many launches, such as one might expect over the decades for a commercial satellite system or a manned military orbiting system, the fully recoverable systems would probably justify their initial development cost. In such high-performance recoverable systems which for a large number of launches will give the lowest cost per launch, one finds, as mentioned above, high-speed air breathing engines using the turbojets and ramjets and high performance liquid rockets; one also notes the entrance of the nuclear rocket into the discussion.

It is important that everything be done that is possible to reduce the high costs of space programs. The United States government alone is spending an amount approaching ten billion dollars on its space research, development, testing, and operations programs—programs which of course are broadly based. The United States program as a whole includes medium- and long-range ballistic missiles, military and commercial satellites, scientific, research, and exploration missions penetrating deeply into the solar system, and scientific measurements of the characteristics of the Earth and the space around the Earth.

New boosters for space missions will cost hundreds of millions of dollars before they can be considered operational vehicles. After they are developed, the operative cost will still remain large until the devices themselves are made recoverable. But the present situation on development and operation is not radically different from that which existed in the development stage of large commercial jet transports. Those also cost hundreds of millions of dollars to develop, but large-scale operational use of jet transports has shown that the effects of the initial cost upon direct operating cost is almost negligible. It follows that it is important in space operation to be able to spread initial development costs over many operations if space is to become an important component of man's everyday life.

Testing: The Booster Experience

As developments in the component fields go forward in a technology as complicated as space flight, there is no substitute for actual space testing, the value of which has been demonstrated by the statistics which come out of the reliability studies on large space boosters.

Thus far in the development of space boosters the percentage of successes in the first ten firings ranges between two out of ten and six out of ten, with the average about four out of ten. From this rough average of forty percent reliability for the first ten shots, reliability goes up into the eighty and ninety percent region as the number of shots increases to a hundred. From there on, increased reliability in boosting gets more and more difficult and seems to be obtainable only with greatly increased numbers of firings.

Another indicator of the improvement that arises from experience is the *Box Score of United States Spacecraft Launches*. A similar box score for Russian launchings over the years is given. Here it is noted that the score is 100 percent for all five years of the space age. This may be explained on the basis of somewhat different rules of scoring.

To back up the engineering developments listed above which are required to improve our space capability, some discoveries in the fundamental sciences of physics, chemistry, mathematics, and biology will be of help. In fact, in the forefront of research in the engineering fields the boundaries between these basic sciences and the engineering fields are fuzzy, and in the modern technological world they tend almost to disappear. This is not true, however, in the design, development, testing, and using of the large, complex space vehicles. Such activities are purely engineering. Here the engineer with purposefulness, vision, and sound technical training will lead the way.

The Nuclear Rocket

As one of the most interesting and widely discussed space developments, the proposed manned Moon exploration is usually publicized in the context of cost. The author has seen estimates of successfully landing a man on the Moon and returning him to Earth ranging from about one billion to one hundred billion dollars, with time estimates from 3 to 20 years. More responsible estimates range from fifteen to forty billion dollars, and from seven to ten years. Granting the importance of cost, in the technical perspective the Moon program is of special interest because it points up the importance of the development of the safe nuclear rocket. Most of the Moon planning programs have been based upon the use of high-performance liquid propellant rockets; and the cost and time estimates of all the development and the testing and the actual missions themselves have been based upon such rockets. But the liquid propellant rocket has been selected only because its principal competitor, the nuclear rocket, is not considered to be in a sufficiently advanced state. If the

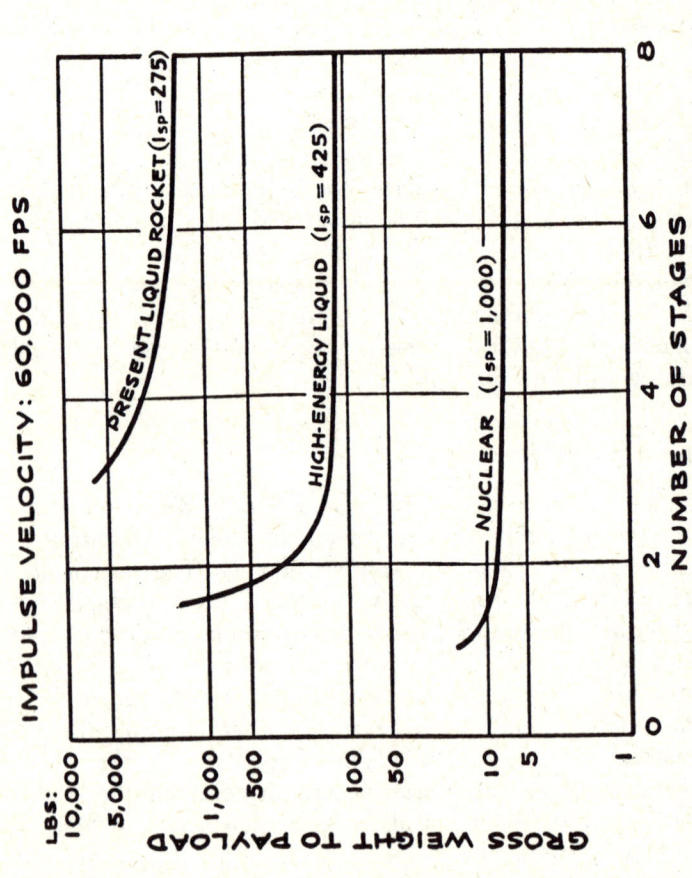

IV-1. Round trip lunar flight with atmospheric braking. From Douglas Aircraft Company.

Moon program objective were moved back to a 15- or 20-year objective instead of something less than ten years, then clearly the nuclear rocket would compete.

Just where, then, does the nuclear rocket stand with respect to its promise for the future and its current development? There has been considerable publicity given to the nuclear rocket development; and in fact a joint National Aeronautics and Space Agency-Atomic Energy Commission development project with industry has already been started. This action follows a long period of experimentation on a test-bed nuclear rocket by the Atomic Energy Commission, aided by certain industries. The new developments are aimed at a flight test engine within a period of five to seven years.

Just how important is the nuclear engine? One answer to this question is suggested by a comparison of liquid, solid, and nuclear propulsion systems in terms of specific impulse, which, measured in seconds, is a merit factor for the efficiency of the use of the propellant.[2] Liquid propulsion systems now are considerably better than solid propulsion systems, and they offer room for further improvement; but the specific impulse promised by nuclear rockets is far above anything that is promised by the liquid or solid rocket propellant systems. It appears that, while the best performance for a liquid propulsion system might be about 500 seconds, and the best for a solid propulsion system about 325 seconds, the best performance of a nuclear propulsion system might be a specific impulse of 1,200 seconds.

Actually, increases in specific impulse multiply over and over in the final performance of specific booster systems. That can best be shown from a graph (*Round trip lunar flight*, etc.). In this graph the ratio of the gross weight at take-off to the weight of the payload is plotted against the number of stages for different kinds of rockets. If one designs a lunar rocket system requiring an impulsive velocity of 60,000 ft/sec, which is quite reasonable, and bases it upon the

[2] Based on a graph, "Propellant Performance," taken from the testimony of Mr. S. K. Hoffman, Vice President of North America Aviation, Inc. and President of their Rocketdyne Division, and a leading developer of rocket engines, to the Committee on Science and Astronautics of the United States House of Representatives.

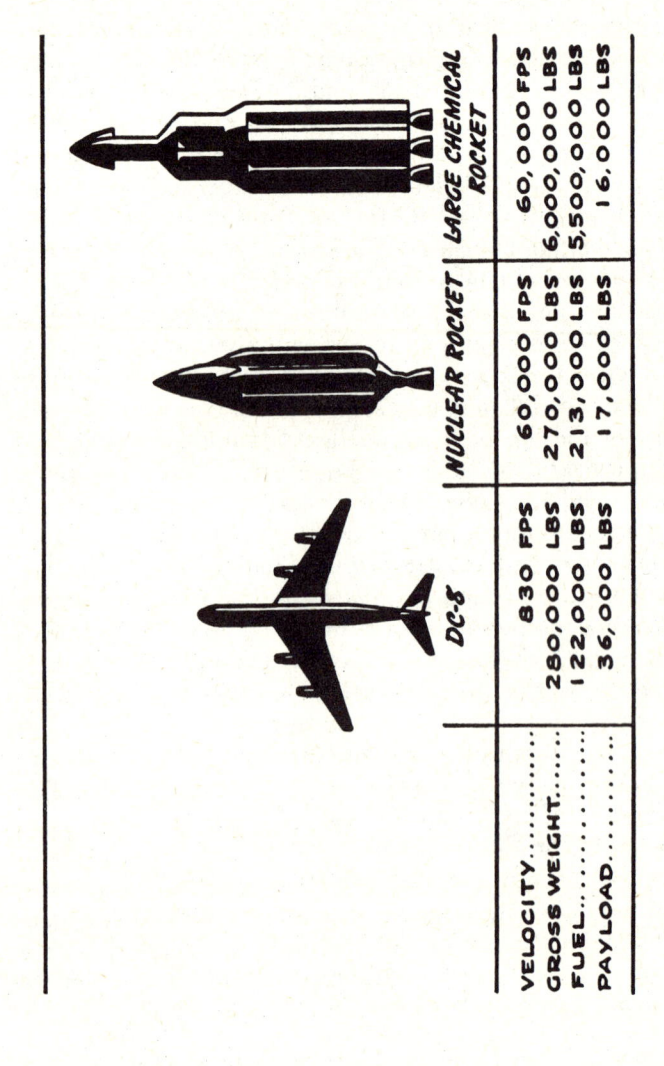

	DC-8	NUCLEAR ROCKET	LARGE CHEMICAL ROCKET
VELOCITY	830 FPS	60,000 FPS	60,000 FPS
GROSS WEIGHT	280,000 LBS	270,000 LBS	6,000,000 LBS
FUEL	122,000 LBS	213,000 LBS	5,500,000 LBS
PAYLOAD	36,000 LBS	17,000 LBS	16,000 LBS

IV-2. Size comparison. From Douglas Aircraft Company.

present liquid rocket specific impulse of 275, a six-stage vehicle would require about 2,000 pounds gross weight at take-off for every pound of payload. If the payload for a manned landing system were 16,000 pounds, the take-off gross weight would be about 32,000,000 pounds. By the time the lunar vehicle was developed, one could count on using high-energy liquid propellant systems with specific impulses possibly as high as 425. The graph shows that gross weight would be something over 100 times the weight of payload. With 150 pounds of payload the number of stages could be reduced possibly to three or four. The same graph shows that the use of a nuclear rocket with a specific impulse of 1,000 seconds would permit the use of about eight pounds of gross weight per payload pound with only two stages.

The reader can get a good idea of the size scale of nuclear and liquid propellant rocket space vehicles by the comparison shown in the accompanying Figure *(Size comparison)*.

Clearly the nuclear rocket offers tremendous promise as a booster system for deep space operations. It is the author's estimate that the development of a satisfactory nuclear rocket-booster system is the *sine qua non* for more distant future space operations. It is also interesting to note that the nuclear rocket does not require new scientific principles. The development of the nuclear rocket will be a long, complicated, expensive engineering project. If, as will almost surely be done someday, the nuclear rocket stage is developed in a form which is recoverable, it will open the door to a reasonable capability for operating in the solar system on rather elaborate missions to distant planets and returning without incurring overwhelmingly exorbitant costs. Such developments will take a long time. The author would not attempt to put a date on the achievement of these final potentialities except to say that it is clearly more than a decade away.

Control, Guidance, and Communications

One whole new field of development consists in the guidance and control equipment for space operations. Ballistic missiles can be guided by two basically different systems: the one employing so-called inertial navigation in which pre-set devices involving gyroscopes and accelerometers and computers guide the vehicle entirely throughout its flight; the other involving radio direction finding and radar distance measuring to perform continuous tracking and guidance of the ballistic missile. For very long-range space missions inertial guidance can be used in part, but there must be corrections to it made by optical or radio sighting devices to reference points such as the Earth, the Sun, and the other planets and stars. A wide range of devices which contribute to this lore has been worked on over the recent decades, and the problem seems to be one not of discovering new principles but of making the technical advances necessary to develop new equipment.

Reliability

In the types of devices needed for communications, guidance, and control there are many electronic, mechanical, and thermal parts which add up to very complex systems. Furthermore, there are restrictions concerning weight and size and power consumption of these parts. Experience with such complicated systems has shown that long-term reliability is always a problem.

Long-term reliability can be obtained for very simple devices. For example, the much maligned Vanguard program, which was hopefully our first but turned out to be later in the series of satellites, and the first of which consisted of only a three-and-a-quarter pound, six-inch diameter sphere with a shell of aluminum containing two very simple radio transmitters, has been operating in space since March 17, 1958. Since the orbit it attained is sufficiently high so that the drag of the Earth's atmosphere does not tend to slow it down, the vehicle is expected to orbit the Earth for a long time, possibly centuries. One of its two transmitters was still broadcasting its position after three years—but it must be understood that this radio transmitter is the simplest of devices. The problem lies in the fact that reliability tends downward rapidly with increased complexity, and that thus far the process of making reliable the complex guidance, control, communication, and other complicated equipment for spacecraft is difficult.

Some indications of the problems of reliability of communications equipment are shown by the tracking of the Sun satellites which have been established by both the Soviet Union and the United States. For example, the Pioneer V had the record interplanetary distance radio communication of something over 22,000,000 miles, before its communications equipment failed. The late Soviet Venus probe failed to transmit after going only a fraction of that distance.

There is one school of thought which believes that the best way of handling the complicated devices for space flight to ensure reliability is to have trained men aboard the spacecraft to repair them. This argument is advanced as one of the most important reasons for putting men into space. In the author's estimate, reliability of space equipment cannot be attained by repair and maintenance operations. One of the biggest problems yet to be conquered in space, it can be solved only by the slow process of improvement of design.

The Space Environment

Whenever man has contemplated going into a new environment—sailing far from his native shores, first going up in balloons and airplanes—he has been challenged by the difficulties, real and imagined, of the new world he is entering. And so with space. For the space environment differs from our environment here on the surface of the Earth. There is no atmosphere to shield humans and equipment from the physical bodies, space particles, and electromagnetic radiations in space; there is no atmosphere to supply life-giving oxygen.

Some of the harmful radiations such as the ultraviolet radiation from the Sun which would burn skin and eyes seriously can be easily shielded with only a small amount of material, such as the structural skin which any spacecraft requires. There are, however, other radiations from the Sun which are much more dangerous and occur mainly in solar flares. For example, there is an extremely high-energy flare from the Sun that might occur, say, once every four years, in which the energy of the particles range in the 370,000,000 electron-volt range which is extremely dangerous to humans and from which they would definitely have to be shielded unless one wanted to take the chance that no human would be exposed when such a flare occurred. There are other solar flares which occur once a month or so and give out heavy radiation, concentrated around 46,000,000 electron volts. These are not as intense but still have to be taken into account. Around the Earth there is a belt of charged particles called the Van Allen Belt. The energies of the particles concentrate around 144,000,000 electron volts and also are of sufficient number that they must be taken into account in shielding. From outside the solar system, from galactic sources, there are cosmic rays with extremely high individual particle energies—around 4,000,000,000 electron volts—but which come in smaller numbers than the others. Finally, if the spacecraft employs a nuclear rocket, there must obviously be shielding for the direct and scattered neutrons and direct and scattered gamma rays from the nuclear reactor.

Throughout the solar system there are very fine particles of dust called micrometeorites, and, scattered in much smaller number, particles larger than dust. The distribution of these particles indicates that there will be some problem due to the slow weathering of outer surfaces of space vehicles by the impingement of this dust, which has a sandblasting effect. Collision of a spacecraft with larger particles would create a hole, but self-sealing techniques and design of multiple-layer skins can minimize this hazard. Possibly the best way to describe the engineering problems raised by the foreign environment of space would be to list the various weights required in a typical vehicle design for a three-man spacecraft intended to travel extensively through the solar system and having a total weight of 52,000 pounds. (This vehicle has not yet been developed, but reasonably complete engineering studies have been made of the system.) The following figures, taken from Douglas Aircraft Company reports, are in pounds:

>   Pressurization and Oxygen System, 630
>   Thermal Conditioning, 720
>   Atmospheric Control, 340
>   Space Suits, 270
>   Three Men, 600
>   Interior Equipment, 560
>   Earth Survival Pack, 234
>   Food and Water, 348
>   Structure, 2,000
>   Shielding, 10,000
>   Electronic Equipment, 1,022
>   Power Supplies, 1,320
>   Last Stage, 20,000
>   Cargo, 14,000

One can see from this summary that the items required to provide the proper thermal and atmospheric control for men are relatively small; even the food and water become small items. The big items are shielding, any cargo or equipment necessary for the men to take on a trip through the solar system, and last-stage propulsion devices. But such a total can be handled by the very large liquid propellant rocket systems and the nuclear rocket systems under development.

Prospects

All that has been said here is quite independent of the special features of national technology, or of the Soviet-United States competition in space. It is clear that Soviet technology has been able to accomplish space missions with boosters larger than those used so far by the United States. However, long-term progress for the Soviets no less than the West will depend on the same considerations spelled out in this chapter.

It is this author's conclusion that, although the problems are many, the currently contemplated space missions are technically possible, and even the hazardous new environment of outer space presents no conditions which are impossible to counter by modern technology. True, the development of all the equipment for providing safe flight for humans in space will be an expensive development program, but on the other hand it seems to be a reasonably straightforward one.

In the end, achievement of the capability to use space profitably for mankind will depend upon the slow, expensive accumulation of engineering experience, not on spectacular breakthroughs in the realm of scientific principles.